Preface

The idea for this monograph originated from a series of lectures on thermoreception given at the University of London. I wanted to consider various aspects of thermoreception including temperature sensation, neurophysiology of thermoreception, thermal comfort and behaviour as well as autonomic temperature regulation. Being well aware that this vast field can only be covered at the expense of comprehensiveness and balance, I have tried to concentrate my presentation on the situation in man. The selection of observations in animals is rather arbitrary, the criterion being the general importance of the results or their possible predictive value for human thermo-physiology.

For preparing the manuscript and the illustrations, I am indebted to Mrs. H. Bestgen, M. Diebel, H. Schenk and M. Thumberger.

January 1981

HERBERT HENSEL

Contents

1
General Concept of Thermoreception

The concept of thermoreception was originally introduced in order to account for the fact that biological thermal sensors are not only involved in conscious temperature sensations but play also an important role for autonomic and behavioural responses of the organism to thermal environments (Hensel, 1952). In 1974, the Encyclopaedia Britannica adopted the term "Thermoreception" (Hensel, 1974a), defining it "as a process in which different levels of heat energy (temperatures) are detected by living things". In this definition, the decisive parameter is temperature as defined by physics.

A genuine approach to thermoreception is based on immediate and specific sensory experiences, such as the qualities "warm" and "cold". From an epistemological point of view, sensory experiences or sensory phenomena are primary elements of knowledge that cannot be reduced to or explained by positive sciences, because such an attempt would already presuppose some other kind of sensory experience. Sensory phenomena are described and analysed by phenomenology, the concepts of which are derived from the qualitative realm of sensory experience *per se*. According to Peirce (1960), phenomenology is "the most primal of all the positive sciences. That is, it is not based, as to its principles, upon any other positive science."

Our view of everyday life as well as the intention of positive sciences, e.g. physics or physiology, is directed towards an objective world in which sensory qualities are considered as properties of objective things. Phenomenology, in its turn, sets out to reduce these more or less complicated conceptual structures to purely phenomenal elements, thereby avoiding *a priori* statements such as "objective" or "subjective". This approach has been described as "analytical reduction" (Reenpää, 1962; Hensel, 1966).

The phenomenal manifold can be structurized by the criteria of similarity and dissimilarity, or by the criteria of dependence or

1

independence. For instance, thermal sensations have something in common but are quite different from the qualities of vision or smell. Furthermore, the quality of warmth can be varied independently of the qualities of colour or weight; thus an object can be warm, green and heavy at the same time but not warm and cold.

Figure 1.1 summarizes the general approaches to thermoreception, including various correlations between sensory phenomena and positive sciences. The phenomenal manifold contains immediate sensory,

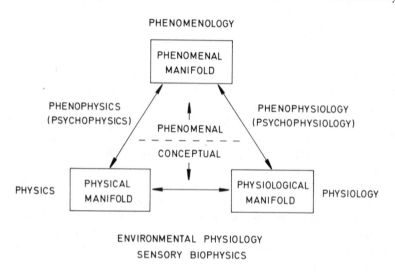

FIG. 1.1. Approaches to thermoreception. For further explanation see text.

affective and volitive experiences (e.g. thermal sensations) described in terms of phenomenology, the physical manifold includes environmental events (e.g. temperatures) described in terms of physics or chemistry, while the physiological manifold contains physiological events (e.g. neural processes) described in terms of physiology, morphology, biophysics, biochemistry and behavioural sciences. Thermal sensory qualities are restricted to the phenomenal manifold, whereas the physical and physiological manifold are dealing with non-thermal observations (Hensel, 1974b). Whereas the content of the first manifold is phenomenal in nature, the contents of the other manifolds are conceptual objects of positive sciences.

The correlation between sensory, phenomenal, and environmental events is treated by phenophysics (or, in classical terms, "psychophysics"), while the correlation between sensory phenomena and physiological events belongs to the field of phenophysiology (or "psychophysiology"). Whereas classical sensory physiology established a causal relationship between "stimulus", "excitation" and "sensation", modern sensory physiology considers the correlation between the phenomenal and the conceptual manifold as a probability implication (Reenpää, 1962; Hensel, 1966, 1973a). For this implication the symbol → is used. The probability implication can be defined as follows: If, according to a certain criterion, an equality group can be formed for the elements (e) of a manifold (A), there is a certain probability (P) that an equality group can also be formed for the elements of a manifold (B).

$$A_{e,1 \ldots e,n} \underset{P}{\rightarrow} B_{e,1 \ldots e,n} \tag{1}$$

$$0 \ldots 1$$

For the implication between phenomenal and physical parameters, a value of $P \approx 1$ means that the physical parameter is an adequate stimulus; in case of $P \ll 1$ one speaks of an arbitrary stimulus.

In phenophysical and phenophysiological experiments, several sensory criteria are used (Hensel, 1966): (1) The absolute threshold, defined as the just noticeable sensation when starting from a zero level of this sensation. In the following, the term "threshold" is used for absolute threshold. (2) The differential or change threshold, defined as the just noticeable change in intensity or quality of an already existing sensation. (3) The perceived intensity or magnitude of a sensation above threshold, usually denoted by an estimated number on a phenomenal magnitude scale.

Various disciplines are dealing with the correlation between environmental and physiological events. The so-called objective sensory physiology describes the relationship between physical or chemical events and processes in receptors and sensory pathways without referring to sensation. Other scientific approaches are concerned with autonomic and behavioural responses of the organism to environmental factors. While the phenomenal manifold is restricted to direct observations in man, the physiological manifold can be investigated both in humans and animals. In the field of thermoreception and temperature regulation considerable physiological differences may exist between various species of mammals as well as between animals and man. As far as the situation in humans is concerned, any analogy from animal experiments has at best the character of a working hypothesis that requires direct testing in man.

Until the beginning of this century, thermoreception was thought to be involved only in conscious temperature sensation. The biological significance of thermoreception in connection with autonomic temperature regulation was neglected, if not explicitly denied. Thus von Frey (1910) wrote: "As yet, no experimentally founded statements can be made about the connections between temperature sense and thermoregulation. . . . It is not attractive to assume a sensory system to be involved in functions that are beyond the limits of consciousness." Another obstacle for realizing the biological function of thermoreception was Weber's theory (1846), according to which the temperature sense responds only to temperature changes but not to constant temperatures. As Thauer (1939) rightly says, this theory is incompatible with the facts of thermoregulation.

2
Principles of Temperature Regulation

2.1 Homeostasis and Variation

Homeothermic vertebrates are able to regulate their internal temperature with a certain degree of accuracy under changing external and internal conditions. Homeothermy is part of the ability of organisms to maintain homeostasis or stability (Ashby, 1966). According to a 1972 glossary by the Thermal Physiology Commission of the International Union of Physiological Sciences, homeothermy is defined as a pattern of temperature regulation in which the cyclic variation in core temperature, either nychthemerally or seasonally, is maintained within arbitrary limits of $\pm 2°$ despite much larger variations in ambient temperature.

There is no clear-cut distinction between homeothermy and poikilothermy. Various poikilothermic species are able to regulate their body temperature by behavioural means under certain conditions for limited periods of time. For example, the "heliotherms" among the reptiles regulate by shuttling between sun and shade but their body temperature can vary by 20 degrees and more with changing internal and external conditions. Thus reptiles may abandon behavioural thermoregulation while defending a territory or avoiding a predator (Templeton, 1970).

However, homeostasis is not the only biological principle. Claude Bernard's almost sacrosanct concept of a virtual straight-line constancy had to be revised more recently to a concept with spontaneous and rhythmic variations (Palmer, 1976). The set point of the thermoregulatory system undergoes rhythmic changes with the diurnal and the menstrual cycle. Moreover, it should be emphasized that human behaviour is not always directed towards a thermal state of maximal constancy and maximal comfort. For various reasons we may even deliberately overtax our thermoregulatory system, e.g. during physical exercise, when taking a hot bath or when swimming in

5

cold water. Obviously man does not only seek comfort, he may also seek stress.

2.2 Concept of Regulation

Biological temperature regulation is made possible mainly by specific central and peripheral nervous structures that constantly detect the organism's temperature fluctuations and attempt to keep them in balance by means of appropriate counter measures. Although the interest of physiologists has been focused on nervous control mechanisms of body temperature, it should not be neglected that various unorthodox theories of thermoregulation have been put forward, according to which other processes, such as autoregulation of tissue temperature, may also be involved in thermoregulation (for references see Bligh, 1973). Without denying their possible importance, these processes will not be discussed here, since this book is dealing with temperature regulation under the aspect of thermoreception. In any case, there is no doubt that normal temperature regulation is only possible with a functioning nervous control system.

2.2.1 CYBERNETIC MODELS

Various attempts have been made to describe the thermoregulatory system in terms of cybernetic models (Hammel, 1968, 1970, 1972; Wyndham and Atkins, 1968; Brown and Brengelmann, 1970; Hardy, 1972a,b; Hensel, 1973b; Bligh, 1973, 1979; Hensel et al., 1973; Houdas et al., 1973, 1978; Bligh and Hensel, 1974; Cabanac, 1975; Werner, 1975, 1977a, 1978; Stolwijk and Hardy, 1977; Brück, 1978a,b). On the one hand, these models are certainly of heuristic value and may advance the formation of clear concepts. On the other hand, the model makers have become increasingly critical about the products of their ingenuity. Analogies from technology may be misleading since the principles actually used by control engineers are only part of the theoretical possibilties of control systems. Another difficulty is that a living regulatory system cannot be opened up completely and analysed in detail without irreversibly impairing its function.

A passive body with a constant rate of heat production at a certain ambient temperature (T_a) will always attain an equilibrium between heat production and heat loss without any regulation. In this steady state, the heat flux density (H) in an element of the body surface is constant with time, and so is the temperature field within the body. The internal temperature (T_i) at some point of the body is a function of H, T_a, and of the total resistance (R) or total insulation through which heat flows from the body to the environment

$$T_i = HR + T_a \qquad (2)$$

From Eqn (2) it follows that, for a given R, the value of T_i changes both with H and T_a. If, for example, the ambient temperature is decreased from T_{a2} to T_{a1}, a new steady state will be attained when the internal temperature decreases from T_{i2} to T_{i1}. If H_1 is increased to H_2, the steady-state condition is an increase from T_{i1} to T_{i2}.

These passive deviations of T_i can be diminished if a suitable parameter is actively changed at the same time. For example, a decrease in T_a can be counteracted by an increase in H, or an increase in H can be counteracted by a decrease in R. Such counteractions can be achieved by means of a negative-feedback control system. In Fig. 2.1 (upper part) a simple proportional control system is shown. It consists of a passive system which may be subjected to a thermal disturbance, for example a decrease in ambient temperature (T_a). This disturbance causes a change in the controlled variable, the temperature (T_i) of the passive system. The value of the controlled variable is detected by a sensor, and a corresponding feedback signal (S) is generated. Any decrease of T_i below a certain value ($T_{i,0}$) constitutes a load error signal (C) that evokes a control action, for example the generation of heat (H). The control action counteracts the deviation of the controlled variable. $T_{i,0}$ is the so-called set point of the system.

Figure 2.1A shows the interaction of the four functions of sensor, controller, effector and passive system. As there is only one quadruple of values compatible with all steady-state functions, it is obvious that these values determine the steady state of the control loop. The steady state functions for the external disturbances T_{a2} and T_{a1} are relatively far from the set point. Any value of the control action (H) is correlated to a certain value of T_i according to the equation

$$H = a(T_i - T_{i,0}) \qquad (3)$$

in which a is a proportionality constant defining the gain of the controller. Likewise, any value of the disturbance (T_a) is correlated to a certain value of T_i. Thus, in the case of Fig. 2.1A, the thermal disturbance is large, and so is the corresponding deviation of T_i from the set point.

In Fig. 2.1B, steady states near the set point are shown, corresponding to higher ambient temperatures (T_{a5} and T_{a6}). T_{a6} is the highest ambient temperature at which a quadruple of all four steady-state functions is possible, since T_i is at the set point and the control action (H) is zero. Still higher values of T_a cannot be counteracted by this control system, and the passive system becomes poikilothermic, that is, it follows passively the fluctuations of T_a.

There is some discussion in the literature whether the thermoregulatory system may function without a set point—a model suggested

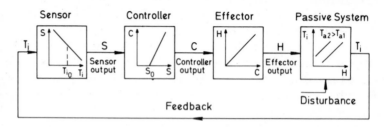

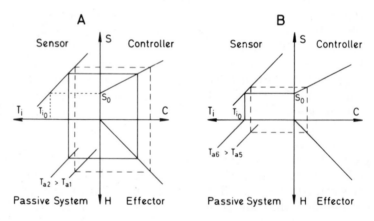

Fig. 2.1. Temperature control system with negative feedback. For further explanation see text. (Modified from Werner, 1978.)

by Werner (1978)—or whether it includes a set point. In fact, both views can be described as two different functional states of the model shown in Fig. 2.1. State A corresponds to Werner's model, and state B to the set point model. The question is, in which state is the thermoregulatory system at "normal" body temperature? As to metabolic and sweating responses, it is in state B, while it may be in state A with regard to vasomotor control.

In contrast to a technical control system, where the set point is *a priori* known by the control engineer, the set point of a living control system must indirectly be assessed. The set point can be defined as that value of the controlled variable at which the control action is zero. In a proportional control system, the control action (y) follows the equation

$$y = a(x - x_0) \qquad (4)$$

where x is the controlled variable, x_0 the set point and a the gain of the controller. From Eqn (4) it follows

$$x_0 = x - \frac{y}{a} \qquad (5)$$

In order to determine the set point, it is thus necessary to measure the response (y) of the effectors and the amount of the controlled variable (x) at the condition $y = 0$. Under this condition x will correspond to the set point.

There are various theoretical possibilities of generating a set point. The feedback signal can be compared with a temperature-independent reference signal, or with a temperature-dependent reference signal of positive or negative slope (Hensel, 1973b). The set point is considered to be at that temperature at which the intensities of both signals are equal. At this temperature the controller output is zero. The set point can be displaced either by a parallel shift or by a change of gain of the temperature intensity characteristics of one signal.

2.3 Is Temperature the Controlled Variable?

Various authors have challenged the view that temperature is the controlled variable. Houdas *et al.* (1973, 1978) suggested a model in which the controlled variable is the heat content (Q) of the body. The load error signal that activates the control actions is a change in heat content (ΔQ). A similar model has been proposed by Webb *et al.* (1978), who assume that the rate of heat outflow is adjusted to metabolic heat production. According to Snellen (1972), there are indications "that the body behaves as if heat regulation is present". Snellen does not maintain that heat content is regulated, because this would require information about the heat capacity of the body. Instead, he puts forward the hypothesis that plasma osmolarity may be a derivative of body volume, and if osmolarity and average body temperature could interact as a constant relationship, heat regulation could be achieved without constancy of body heat content. However, the correlation between osmolarity and body volume holds, at best, for short-term heat stress, but not for ontogenetic or nutritional changes of body volume.

In the model proposed by Houdas *et al.* (1973, 1978), changes in heat content are measured as changes of temperature at various sites by means of signals originating in all the thermoreceptors of the body. What is measured is an average body temperature. As long as the heat capacity of the body does not change, average body temperature goes parallel with heat content. But with changing heat capacity, as it occurs during growth, or during gain or loss in body weight in adults, or after loss of extremities, this correlation does not hold any longer. If heat content were the controlled variable, a growing human being with a 20-fold increase in heat capacity would undergo a drastic decrease in body temperature; but, in fact, the body temperature from birth until the adult age remains surprisingly constant (cf. p. 241. Furthermore, a regulation of heat content could not account for the fact that various mammals have similar body temperatures (for references see Hensel *et al.*, 1973), although the heat content of a whale can be 10^7 times that of a mouse.

Neither would the facts of temperature regulation be compatible with the assumption that the rate of change in heat content (dQ/dt) is the load error signal. If, for example, temperature regulation is overtaxed in the cold and the organism becomes hypothermic, a new thermal equilibrium of the passive system may be reached at subnormal body temperature. A steady state between heat production and heat loss is attained at which dQ/dt becomes zero and consequently no control actions would be present. But, in fact, control actions can still be active during hypothermia. If the hypothermic organism is transferred to a somewhat warmer environment, a certain quantity of heat would be gained and dQ/dt become positive. This, in turn, would activate heat dissipation mechanisms. However, what actually occurs is the activation of heat conservation mechanisms until the body temperature has reached its normal level.

Summing up the available evidence, the conclusion seems justified that the facts of temperature regulation are best described by the theory that temperature—or a function of various temperatures—is the controlled variable (cf. Bligh, 1978).

2.4 Physiological Temperature Regulation

2.4.1 MULTIPLE-LOOP SYSTEMS

The thermoregulatory system of homeotherms is more complicated than any engineer's blueprint; it is grossly non-linear in the mathematical sense and has no single controlled variable and a high redundancy; it contains multiple sensors, multiple feedback loops, and multiple outputs.

Figure 2.2 shows a schematic diagram of autonomic temperature regulation in man. The controlled variable is an integrated value of multiple temperatures rather than the temperatures of a limited area of the body core, e.g. the hypothalamus (Brown and Brengelmann, 1970; Mitchell *et al.* 1970; Bligh, 1972; Snellen, 1972; Hensel, 1973b; Bligh and Hensel, 1974; Brück, 1978a,b). Snellen (1972) has proposed that average body temperature might be the controlled variable.

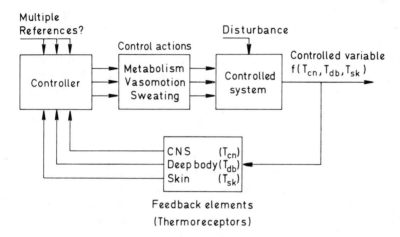

Fig. 2.2. Schematic diagram of autonomic temperature regulation in man. T_{cn}, central nervous temperature; T_{db}, extracentral deep body temperature; T_{sk}, skin temperature.

However, this would require a homogenous distribution of thermoregulatory inputs within the body. In fact, the distribution is highly inhomogenous.

For practical purposes, a "weighted mean body temperature" $(\bar{T}_b)$ may be considered as the controlled variable. In most cases a combined value of internal body temperature (T_i) and average skin temperature $(\bar{T}_s)$ is used for humans according to the equation

$$\bar{T}_b = aT_i + (1 - a)\ \bar{T}_s; a < 1 \qquad (6)$$

When measuring T_i in the oesophagus, values of a have been proposed between 0.9 (Nadel *et al.*, 1971a) and 0.87 (Brück *et al.*, 1976). This weighting ratio between a and $1 - a$ is assumed to be the relative effect of $\bar{T}_s$ and T_i in a linear control function (Gonzalez *et al.*, 1978).

The set temperatures for the different control actions such as heat production, cutaneous vasomotion and sweating need not necessarily be identical. For example, heat-dissipation mechanisms driven by warm sensors may have a higher set temperature than heat-production mechanisms driven by cold sensors. Thus in a certain temperature range a "dead band" or "zone of thermal neutrality", or "interthreshold zone" would occur where no regulation takes place. On the other hand, heat dissipation may have a lower set temperature than heat production, both mechanisms thus being simultaneously active in a certain temperature range without any "dead band". In this case a "virtual" set point can be defined as that temperature—or function of temperatures—at which the net output of the control system is zero. It has further been shown that the set temperature of different regulatory functions may vary separately and independently.

In a control system with a single feedback loop, the loop may be opened in order to assess its "open loop gain", that is, the change in core temperature (ΔT_{co}) divided by the corresponding temperature change at the sensor (ΔT_{se}).

$$\text{Open loop gain} = \Delta T_{co}/\Delta T_{se} \qquad (7)$$

However, in a multi-loop system it is not possible to measure open loop gain because an unknown number of other closed loops will counteract the displacement of temperature by setting up feedback signals. Instead, one can assess the relative sensitivity of a thermore-ceptive site by feedback signals (Jessen and Clough, 1973a), that is, by a static change in core temperature. Of course, this change is not as high as when open loop gain would be measured.

Another possibility of assessing the sensitivity of a thermoreceptive site is to measure the control actions. Here the same objections can be raised as when measuring open loop gain. Under certain conditions it might be possible to measure steady control actions during an early phase before body temperature has markedly changed and thus influenced other feedback loops.

2.4.2 AUTONOMIC AND BEHAVIOURAL CONTROL

The diagram Fig. 2.3 shows some basic features of autonomic temperature regulation and thermoregulatory behaviour in man. Thermal disturbances have two main sources: (1) internal heat generation by exercise and (2) environmental heat or cold. Signals from external temperature disturbances can rapidly be transferred to the central nervous system by means of cutaneous thermoreceptors via afferent nervous pathways before the disturbance has reached the core of the body. Important in this respect is the ability of cutaneous thermoreceptors to respond not only to the temperature (T) but also to the rate of a temperature change (dT/dt). The proportional and rate control is so effective that rapid external cooling or warming may result in a transient opposite change of internal temperature.

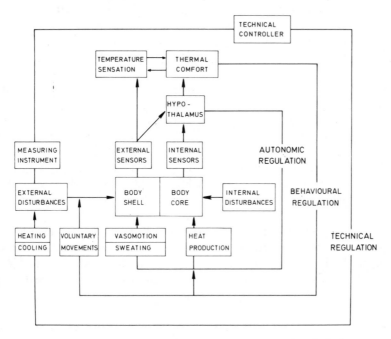

FIG. 2.3. Schematic diagram of autonomic, behavioural and technical temperature regulation in man.

Usually discriminations are made between autonomic and behavioural regulations. However, a sharp distinction has proved impossible. Behavioural responses to heat and cold modify the relations between organism and environment and thereby modify the need for autonomic thermoregulatory responses, but this sparing action does not necessarily imply central nervous coordination between behavioural and autonomic thermoregulation.

In man, behavioural thermoregulation is associated with conscious temperature sensations as well as with emotional feelings of thermal comfort and discomfort (Fig. 2.3). Whereas temperature sensations seem to depend mainly on the activity of cutaneous thermoreceptors, thermal comfort and discomfort reflect a general state of the thermo-regulatory system, or, in neurophysiological terms, an integration of afferent signals from both cutaneous and internal thermoreceptors. Temperature sensations elicited from the skin are used to judge the thermal state of objects or environments and may thus have a predictive value for human behaviour. In a wider sense, the technolo-gical inventions of man for artificial control of his thermal environ-ment can be considered as part of his behavioural thermoregulatory ability. In this case the regulation of temperature is shifted from the body to the environment, with artificial sensors, controllers and effectors.

Table 2.I summarizes various autonomic and behavioural compo-nents of temperature regulation. Their physical importance can be derived from the general laws of stationary heat flux in the organism (for details see Hensel et al., 1973; Gagge and Nishi, 1977). More elaborate mathematical models for the heat transfer of the human body have been developed by various authors (e.g. Brown, 1965; Wissler, 1966; Wyndham and Atkins, 1968; Mitchell et al., 1972; Werner, 1975, 1977a; Hayward et al., 1977; Stolwijk and Hardy, 1977).

Since the body continuously produces heat, a steady state is only possible if there is a continuous heat flux along a temperature gradient from the body core to the environment. The heat flux density (H) in an element of the body surface in the steady state can be written as

$$H = h_i \, (T_i - T_s) \tag{8}$$

where h_i is an internal heat transfer coefficient, T_i the temperature of a point inside the body and T_s the temperature of the surface element. The heat flux density in the same surface element is also

$$H = h_a \, (T_s - T_a) + H_e \tag{9}$$

where h_a is an external heat transfer coefficient, T_a the ambient temperature and H_e the heat absorption by evaporation, expressed as heat flux density. For simplicity it is assumed that the heat flux by conduction, convection and radiation is proportional to the tempera-ture difference. Since the heat flux density from the core to the surface and from the surface to the environment is equal in the steady state, we obtain from Eqns (8) and (9) for T_i

$$T_i = T_a + HR_i + R_a \, (H - H_e) \tag{10}$$

The reciprocal values $1/h_i$ and $1/h_a$ are replaced by the symbols R_i and R_a. They have the meaning of an internal and external thermal

TABLE 2.I

Control actions of temperature regulation in homeotherms

Physical factor	Autonomic regulation	Regulatory behaviour
Ambient temperature T_a		Migration
		Seeking sun or shade, etc.
		Artificial heating and cooling
Heat production H	Shivering	Active movements
	Non-shivering thermogenesis	Food intake: Specific dynamic action
		Warm and cold food
Internal thermal resistance	Cutaneous blood flow	Clothing
including integument R_i	Erection of hair and feathers	
External thermal resistance R_a	Respiration: dry heat loss	Nest building
		Seeking shelter
		Seeking ground surface of
		various thermal conductivity,
		wind, water, etc.
		Air movement by fanning
		Ventilation
Water evaporation H_e	Sweat secretion	Moistening of body surface
	Respiration: evaporative heat loss	with water, saliva or nasal fluid
	Secretion of nasal and oral glands	Moistening of clothing
Geometric factor		Body posture
		Huddling of several individuals

From Hensel *et al.* (1973).

resistance or insulation. Thermal resistance is also dependent on the geometry of the body.

The organism is able to change actively the values of H, R_i and H_e, and to some extent also R_a. Thermal disturbances arise mostly from T_a, R_a and H. It is important to realize that the heat production of the body acts not only as a control action but can also cause thermal disturbances. In man, heat production during extreme work can be many times higher than the basic metabolism. This excess heat must be compensated for by other physiological factors such as sweating.

The aim of regulation is, somewhat simplified, to keep T_i constant. Homeotherms can activate their heat production as well as their heat resistance for temperature regulation. Equal ambient and core temperatures ($T_a = T_i$) are only possible if the sum of the second and third term on the right side of Eqn (10) becomes 0. Since H, R_i and R_a are always positive, only the term $H - H_e$ remains which becomes negative if $H_e > H$. The amount of H_e can further be increased, so that some homeotherms, in particular humans, can even exist under the condition $T_a > T_i$. Formally, heat absorption by evaporation can also be considered as a "negative heat resistance", that is, a resistance which allows heat to flow from a lower (T_i) to a higher (T_a) temperature. Evaporation of water is thus the only means by which body temperature can be maintained at higher ambient temperatures.

2.4.3 DEVIATIONS OF BODY TEMPERATURE

In a proportional control system (Eqn 4), the control action (y) is proportional to the difference between feedback signal (x) and set point (x_0). The larger the disturbance and, therefore, the control action, the larger is the deviation ($x - x_0$) of the controlled variable. If physiological temperature regulation is assumed to function as a proportional control system, any thermal load will necessarily lead to a corresponding deviation of body temperature ("load error"). The higher the gain of the control system, the smaller the load error.

Besides these normal deviations of body temperature, there are various possibilities of changing the temperature of the body: (i) external heat or cold stress overtaxes the thermoregulatory system; in this case the properties of the system are unchanged but the quantitative capacity of the control actions is insufficient; (ii) primary impairment of the controller output, such as paralysis of cutaneous blood vessels, artificial metabolic increases, etc.; (iii) changes or disturbances of the controller itself. An increase in core temperature above a standard range is called hyperthermia, whereas a corresponding decrease is called hypothermia. Fever is a rise of internal temperature due to changes in the central nervous controller in the sense of a shift in set point. A corresponding hypothermia has no special term; it may be called "central hypothermia".

2.4.4 INTERFERENCE OF REGULATIONS

There is considerable interference between thermoregulation and other biological control systems. In many cases we find loops with two or more regulatory systems sharing a common output with thermo-regulation, e.g. energy production, water metabolism or cutaneous circulation. This involves the possibility of competition between various systems. In many instances thermoregulation turns out to be very powerful. Homeotherms living in the cold may keep their body temperature at high levels even during starvation, instead of saving energy by hypothermia. Temperature regulation may also interfere with water metabolism, as is observed during heat where sweat secretion is maintained at high rates even when the body becomes dehydrated. Another example is the failure in blood pressure regulation due to excessive vasodilatation in the heat (heat syncope). In other instances, however, the narrow limits of homeothermy are abandoned, thereby saving food or water, as true for hibernators in the cold and for camels and antelopes in hot climates (for references see Hensel et al., 1973).

3
Temperature Sensation in Man

3.1 Structure of Temperature Sensation

It is relatively easy to discriminate the phenomenal qualities of warm and cold from the manifold of cutaneous sensations. Both qualities form a sensory continuum of various intensities:

> painful (cold)
> icy
> cold
> cool
> neutral or indifferent
> lukewarm
> warm
> hot
> painful (hot).

Whether the sensation of "heat" is only a more intense warm sensation or a mixture of various qualities is not quite clear. According to Alrutz (1898, 1900) heat is a combination of warmth and "paradoxical" cold. The sensory quality brought about by simultaneous application of warm and cold stimuli was interpreted by several authors as "heat" but statistical investigations in untrained subjects have shown that the number of "heat" judgments decreased when cold stimuli were added to pure warm stimuli (for references see Hensel, 1952a). A quality of "heat" was sometimes reported by uninstructed subjects when a cold centre stimulus was bracketed by warm outer stimuli (Green, 1977). Perhaps "heat" might be considered as a quality of its own.

Similar problems arise on the side of more extreme cold sensations. It is doubtful whether the sensation of "icy" is only a higher intensity of "cold" or a change in quality, as spontaneously reported by

18

untrained subjects (Beste and Hensel, 1977; Beste, 1977). Erickson and Poulos (1973) have put forward the hypothesis that the scale of thermal sensations is not only a matter of intensity but also of quality.

Towards the extremes of the intensity scale, warm and cold sensations become increasingly painful. Finally, there remains heat and cold pain, while the intensity of specific thermal sensations diminishes. Heat and cold pain belong to the realm of nociception rather than that of thermoreception.

Thermal comfort and discomfort are "pleasant" and "unpleasant" emotional feelings which can phenomenologically be discerned from temperature sensations. Physiologically, thermal comfort and discomfort reflect an integrated state of the thermoregulatory systems. This means that not only cutaneous receptors but also thermosensitive structures in the central nervous system and other parts of the body are involved.

3.2 Physical Conditions of Temperature Sensation

Although the physical concept of temperature has some connection with thermal sensations, the physical temperature scale is based on arbitrary definitions. Therefore it cannot be expected that any simple correlation exists between phenomenal and physical temperature. The most striking difference is that the phenomenal scale has a dual or polar character, namely, "warm" and "cold", separated by a zone of indifference, whereas the physical temperature scale has a monotonic structure, the only quality being warmth.

As a result of numerous investigations, the general physical correlates of threshold and intensity of warm and cold sensations (E_T) can be expressed as a function of (1) the absolute temperature (T) of the skin, (2) the rate of change (dT/dt) of skin temperature and (3) the stimulus area (F).

$$E_T \rightarrow f(T, \, dT/dt, \, F) \qquad (11)$$

Because of their intracutaneous site, thermoreceptors have neither the temperature of the skin surface nor that of the blood. Any reliable metrics of thermal stimuli has thus to account for the temperature in different layers of the skin. By means of fine thermocouples it has been possible to measure directly the intracutaneous temperature field under static and dynamic conditions and to calculate the thermal diffusion coefficient of the skin, which turned out to be 5 to $13.10^{-4} \, \mathrm{cm^2 \, s^{-1}}$ (Hensel, 1952a).

3.2.1 TEMPERATURE AND TEMPERATURE CHANGES

When a small cutaneous area (e.g. 20 cm^2) is adapted to a temperature of 34 °C, the subject will feel neither warm nor cold. Linear
temperature rises from this point of indifference lead to warm
sensations, linear cooling to cold sensations. The threshold (ΔT) of
warm or cold sensations deviates the more from this point, the slower
the temperature is changed. By plotting the rate of change (dT/dt)
versus the thermal threshold (ΔT), a hyperbolic function is obtained
(Fig. 3.1). Similar results have been found by Kenshalo *et al.* (1968).

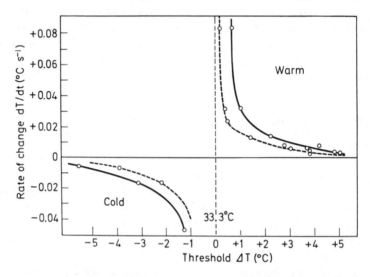

FIG. 3.1. Average thresholds (ΔT) of warm and cold sensations on the forearm
(20 cm^2). Dashed lines thresholds; solid lines distinct sensations. (From Hensel,
1952a.)

At high and low temperatures, the threshold rate of change finally
reaches zero which means that steady temperature sensations occur at
constant skin temperatures. According to the areal extent and the
particular region of the body stimulated, the limits may be found at
different temperatures. For example, with controlled intracutaneous
temperature and 20 cm^2 stimulus area on the forearm, the limit for
steady cold sensations was 20 °C and for steady warm sensations
40 °C (Hensel, 1952a).

Starting from various adapting temperatures, the threshold (ΔT)
for warm sensations at equal rates of warming increases with decreas-

ing adapting temperature. In the diagram Fig. 3.2 the warm
thresholds (ΔT) are depicted as a function of adapting temperature
(T) for various rates of stimulus temperature changes. An analogous
behaviour has been found for cold sensations. The fact that the warm
thresholds increase at lower adapting temperatures, while the highest
cold thresholds are found at high adapting temperatures is in good
agreement with numerous investigations (for references see Hensel,
1973a) and also with the observation of daily life that we feel less cold
when the body was well warmed up previous to cooling.

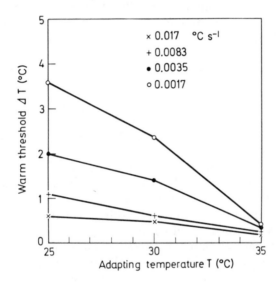

FIG. 3.2. Average thresholds (ΔT) of warm sensation on the hand as a function of
adapting temperature (T) for linear temperature rises from 0.0017 to 0.017 °C S⁻.
(From Hensel, 1952a.)

Besides the measurement of warm and cold thresholds, numerous
investigations have been carried out in which the intensity of warm
and cold sensations were estimated in a magnitude scale; most of
these investigations were based on the so-called "psychophysical
power law" postulated by Stevens (1958, 1960, 1970). However, some
doubts must be raised against the general validity of this function, in
particular as temperature sensation is concerned. (1) The power law is
based on the assumption that the scale of estimated magnitude is a
ratio scale, which means that the axioms of multiplication (commuta-
tivity, associativity and distributivity) should be valid. To my know-

ledge, this has not yet been tested experimentally for the temperature sense. (2) The type of correlation between estimated magnitude and stimulus depends on the arbitrary definition of the stimulus. For example, the estimated magnitude of a constant temperature sensation is not a power function of temperature. (3) The statement that neural events in sensory systems are power functions of the stimulus (Stevens, 1970) has only limited validity. The static discharge of peripheral cold receptors is a bell-shaped function of static temperature, and the same holds for the neural activity in central thermosensitive pathways (cf. Chapters 4 and 5). For dynamic conditions the correlation between estimated magnitude of thermal sensation and temperature change can more satisfactorily be described by power functions.

Using rapid warm and cold stimuli ($16\,cm^2$) on the forearm, apparent warmth and cold were found to grow as power functions of the difference between the neutral adapting temperature ($32.5\,°C$) and the stimulus temperature, the exponent being $n = 1.6$ for warmth and $n = 1.0$ for cold (Stevens and Stevens, 1960). Starting from adapting temperatures of $34\,°C$, cooling the skin with temperature transients of approximately rectangular shape elicited a cold sensation, the estimated magnitude of which was a linear function of the magnitude of the cold stimulus over a range of $10\,°C$. Four categories of magnitude could be correctly identified over this range with 95% confidence. Warming the skin from $34\,°C$ with rectangular stimuli revealed a similar relationship between sensation and stimulus, the number of categories being five over the $10\,°C$ range (Darian-Smith and Dykes, 1971). With more prolonged warm stimuli (heat radiation), perceived warmth grew as a power function of the difference between the irradiant flux of the stimulus and the flux that approximates the absolute threshold for warmth. However, from the shortest to the longest stimulus durations (12 s), the exponent of the power function rose from $n = 0.87$ to $n = 1.04$. Neither the change in tissue temperature nor its rate of change, nor the spatial intracutaneous temperature gradient correlated consistently with the magnitude of perceived warmth. The findings suggest that sensory adaptation might act in such a way as to nearly offset the effect of rising skin temperature with increased duration of stimulation (Marks and Stevens, 1968). As has been pointed out earlier (Hensel, 1952a), the term "adaptation" is the logical equivalent for the fact that time acts as a stimulus parameter.

Johnson *et al.* (1979) have estimated difference limens for paired warming pulses of nearly rectangular shape applied to the palm of the

hand by a stimulator of 1.1 cm². Starting from an adapting temperature of 34 °C, the difference limen, as defined by a 0.75 probability of
correct judgments, varied between 0.03 and 0.06 °C in trained
subjects for stimulus magnitudes between 0 and 8 °C, the lowest
difference limen occurring at stimulus magnitudes around 5 °C. The
correlation between these measurements and recordings of afferent
impulses from warm fibres in the palm of monkeys is discussed in
Chapter 6.

When warm and cold stimuli (area 14 cm²) of 1, 2 and 5 °C
magnitude and 0.4, 1 and 5 °C s⁻¹ rate of change were presented at
adapting temperatures of 35 °C, the estimated magnitude of sensation

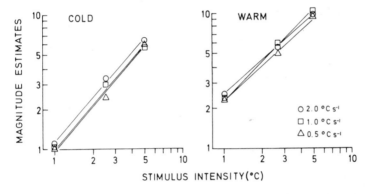

FIG. 3.3. Estimated magnitude of thermal stimuli (area 14 cm²) of various intensity
and various rate of change. (From Molinari *et al.*, 1977.)

was found to follow a power function of stimulus magnitude, but to be
practically independent of the rate of change (Fig. 3.3). For warm
stimuli, the exponent of the power function varied between $n = 0.86$
and 0.95 with various rates, while the corresponding values for cold
stimuli were $n = 1.08$ to 1.14 (Kenshalo, 1976; Molinari *et al.*, 1977).

In contrast to investigations under dynamic conditions, there are
only a few experiments with steady temperature sensations and
constant temperatures. The diagram Fig. 3.4 shows the estimated
static temperatures of the hand resting on a metal thermode of 75 cm²
kept at constant temperatures for 30 min. The estimation was made
both in degrees Celsius and in relative values on a scale between 40
and 25 °C. Under these conditions, there is practically no temperature
range at which complete adaptation of thermal sensation occurs. The
estimation in °C is most correct at 37 °C, while the temperatures

above this level are estimated to be higher than the physical value in degree Celsius and the temperatures below 37 °C are estimated to be lower than the corresponding physical value. Below 27 °C, the curve of temperature estimation becomes practically horizontal and thus the static difference limen relatively large (Beste and Hensel, 1977; Beste, 1977).

The estimated magnitude of temperature difference between the left hand continuously immersed in water of 28 °C and the right hand

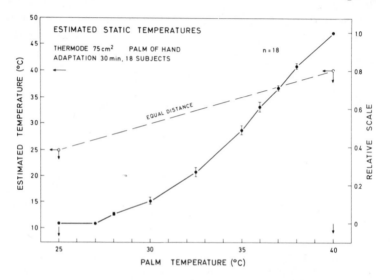

FIG. 3.4. Estimated static temperatures as function of constant temperatures of the palm. Left scale: estimates in °C. Right scale: estimates in relative units. "Equal-distance" line indicates identity between estimated and physical temperatures in °C. Average values of 18 subjects; bars indicate standard error of mean. (From Beste, 1977.)

exposed to constant temperatures between 28 and 17 °C was practically zero until the temperature of the right hand was set below 23 °C (Fig. 3.5). Then the hand was clearly judged to be cooler than the other one but, as a number of subjects spontaneously remarked, this was mainly achieved by a change in quality of sensation becoming more "icy" and gradually painful (Beste and Hensel, 1977; Beste, 1977). It is clear that the relative sensitivity of static temperature sensation cannot be expressed by a power function of temperature. The measurements by Erickson and Poulos (1973) of static difference limens between the left and right hand, although obtained with different methods, bear at least some qualitative similarity with these results, in that the discrimination was least precise between 29 and 26 °C and increased in precision below 25 °C.

3.2.2 SPATIAL TEMPERATURE GRADIENTS

Stimulation of thermoreceptors by intravenous injection of cold and warm solutions led to the same sensations as when the receptors were

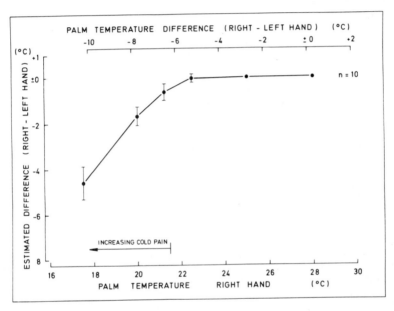

FIG. 3.5. Estimated difference in °C between left hand and right hand immersed in flowing water of constant temperature. Left hand was continuously kept at palm temperature of 28 °C, right hand exposed for 30 min to palm temperatures indicated on the abscissa. Average values of 10 subjects; bars indicate standard error of mean. (From Beste, 1977.)

stimulated from the skin surface (Hensel, 1952a). Thermoelectric measurements at various depths (x) of the skin have shown that on intravenous cooling and warming, the spatial temperature gradient (dT/dx) in the skin is reversed in comparison with the gradient on stimulation from the skin surface. Thus the general condition for warm and cold sensations is warming or cooling *per se*, independent of direction and slope of the intracutaneous spatial temperature gradient. Investigations with microwave and infrared heating of the skin have led to similar results (Vendrik and Vos, 1958).

3.2.3 STIMULUS AREA

Numerous investigations have revealed a considerable influence of stimulus area on threshold and intensities of temperature sensation (Weber, 1846; Hardy and Oppel, 1937, 1938; Hensel, 1973a; Stevens *et al.*, 1974; Kenshalo, 1976). The warm threshold at uniform rates of linear temperature increases may vary by several degrees when the stimulus area is changed between 1 and 1000 cm^2 (Fig. 3.6). It should be mentioned here that the results from stimulus areas less than 1 cm^2 are difficult to interpret, since the factor of three-dimensional heat

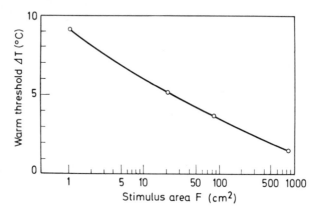

FIG. 3.6. Average thresholds (ΔT) of warm sensations on the forearm for linear temperature rises of 0.017 °C s^{-1} as a function of stimulus area (F). Initial temperature 30 °C. (From Hensel, 1952a.)

flow becomes decisive at small areas. Thus the dependence of sensations on stimulus area might be a physical rather than a physiological effect (Hensel, 1952a).

The estimated magnitude of a dynamic warm stimulus depends on stimulus area as well as on the particular region of the body stimulated. When applying warm stimuli between 4.7 and 200 cm^2 to the back (Fig. 3.7), the exponents of the power functions varied between $n = 2.00$ and 0.67 with increasing stimulus area (Stevens *et al.*, 1974). Figure 3.8 shows the estimated magnitude of warmth as function of stimulus area for different regions of the body.

Spatial summation occurs not only on a single site of increasing area, but also when two sites, symmetrically located on opposite sides of the body, are simultaneously stimulated. Hardy and Oppel (1937)

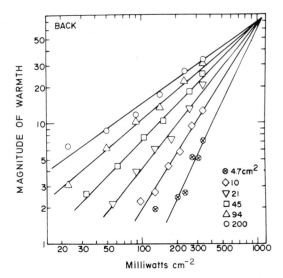

FIG. 3.7. Magnitude estimates of warmth as a function of radiant intensity applied to the back. The parameter is the areal extent of stimulation. (From Stevens *et al.*, 1974.)

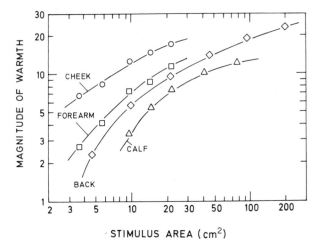

FIG. 3.8. Magnitude estimates of warmth as a function of area for a stimulus of 200 mW cm^{-2} applied to four body regions. (From Stevens *et al.*, 1974.)

found that the threshold for cold stimuli was lower when the backs of the two hands were simultaneously stimulated than when either hand was stimulated alone. Spatial summation was not observed when the forehead and the contralateral hand were simultaneously stimulated.

Similar observations have been made for sites on either side of the midline of the forehead (Marks and Stevens, 1973) and between dermatomes on the back (Banks, 1973).

With symmetrically located thermodes of 18.4 cm^2 on both forearms, the threshold for cold stimuli was significantly lower when both sides were stimulated simultaneously than when either side was stimulated alone (Rózsa and Kenshalo, 1977). The data indicated complete spatial summation (area and stimulus magnitude trade reciprocally) for cold stimuli near threshold and at clearly suprathreshold intensities. This spatial summation is invariant of the adapted skin temperature at which the cold stimuli started. Rózsa and Kenshalo (1977) assume that the underlying neural integration of bilaterally symmetrical thermal afferents may occur in the spinal cord by way of transcommissural interneurons. This would perhaps explain that this type of summation apparently does not occur in the same way when combining non-symmetrical thermal stimuli applied at different sites of the body.

In connection with the biological importance of thermoreception in man, the results obtained by stimulation of large cutaneous areas are of particular interest. When warming the whole body in a climatic chamber with linear temperature rises, even extremely slow rates of change of 0.001 °C s^{-1} are sufficient to elicit warm sensations at skin temperatures of 35 °C (Maréchaux and Schäfer, 1949).

In everyday life, the need for accurate spatial discrimination of thermal sensations is minimal. Further, most "hand size" objects have a relatively uniform temperature over their surface that eliminates the need for spatial discrimination of warmth or cold. There is some interference between tactile and thermal sensations, in that the latter are referred to a site of nearby tactile stimuli (Green, 1977, 1978). On the arm and fingers, warmth refers more strongly to a tactile stimulus than does cold, and referral of warmth is affected relatively little by increasing the distance between stimulators (Green, 1978).

3.3 Regional Differences in Thermal Sensitivity

3.3.1 COLD AND WARM SPOTS

Numerous authors have described the distribution of cold and warm spots in the skin of man, using temperature sensation as a criterion. In general, cold spots seem to be distributed more densely than warm spots (Table 3.I). References for the topography of cold and warm spots in human skin are given by von Skramlik (1937) and Hensel (1952a).

From experiments in human subjects alone it is difficult to decide whether the temperature spots represent discontinuous cold and warm sensitive fields in an otherwise thermally insensitive area or whether they are maxima in a thermosensitive continuum. Particular difficulties arise from the use of spot-like stimuli (cf. p. 26), and it

TABLE 3.I

Number of cold and warm spots per square centimetre in human skin

	Cold spots[a]	Warm spots[b]
Forehead	5.5–8.0	
Nose	8.0	1.0
Lips	16.0–19.0	
Other parts of face	8.5–9.0	1.7
Chest	9.0–10.2	0.3
Abdomen	8.0–12.5	
Back	7.8	
Upper arm	5.0–6.5	
Forearm	6.0–7.5	0.3–0.4
Back of hand	7.4	0.5
Palm of hand	1.0–5.0	0.4
Finger dorsal	7.0–9.0	1.7
Finger volar	2.0–4.0	1.6
Thigh	4.5–5.2	0.4
Calf	4.3–5.7	
Back of foot	5.6	
Sole of foot	3.4	

[a] From Strughold and Porz (1931).
[b] From Rein (1925b).

cannot be excluded that the response of a single thermoreceptor might be below the threshold of conscious sensation, whereas larger stimulators would lead to thermal sensations because of spatial summation. This possibility should especially be considered in case of the warm spots, the density of which, as established by phenophysical methods, seems extremely low in certain skin areas. Of course, a single cold or warm spot must not necessarily correspond to a single thermoreceptor established by neurophysiological methods.

3.3.2 REGIONAL THERMOSENSITIVITY

An important question in connection with temperature sensation, autonomic temperature regulation, thermal comfort and thermoregulatory behaviour is the regional distribution of thermosensitivity in the skin. If thermal sensitivity differs in various regions of the body, these regions may be of different importance for eliciting sensory or regulatory responses, even when the surface areas are equal.

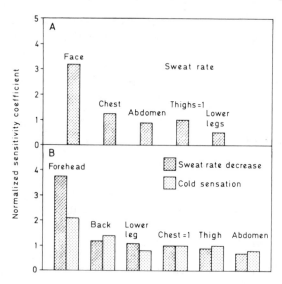

Fig. 3.9. Sensitivity coefficients of various body areas. A: Sensitivity to warming assessed by reflex sweat rate. (From Nadel *et al.*, 1973.) B: Sensitivity to cooling assessed by decrease in reflex sweat rate and cold sensation. (From Crawshaw *et al.*, 1975.)

In human subjects, this problem was investigated by applying equal temperature stimuli to various skin areas and following up autonomic thermoregulatory responses as well as thermal sensations. According to Stevens *et al.* (1974), the warm sensation caused by equal heat irradiation of the skin descended in the following order: forehead, chest, abdomen, back, thigh and calf. At high levels of irradiation, these differences were less pronounced.

When different areas of the human body were exposed to equal intensities of heat irradiation, the sweat rate measured at the skin of the thigh varied with the stimulated region (Nadel *et al.*, 1973; Crawshaw *et al.*, 1975). Thermal sensitivity in coefficients for sweating were obtained as the increase in thigh sweating rate divided by the area irradiated and the increase in skin temperature. The normalized sensitivity coefficients (thigh coefficient = 1) are shown in Fig. 3.9A. The sequence of sensitivities for sweating was of the same descending order as found for warm sensation.

Similar experiments were performed with cold stimuli applied to various skin areas by a water-cooled thermode at ambient temperatures of 39 °C (Crawshaw et al., 1975). Criteria were the decrease in

TABLE 3.II

Mean skin temperature ($\bar{T}_{sk}$) weighting factors as determined by weighting for area, for area and sensitivity of local sweating rate to warming, and for area and sensitivity of local sweating rate to cooling. Since data on sensitivity to cooling is not available for the upper and lower arms, these regions were given simple area weightings.

	$\bar{T}_{sk}$ (area weighting only)	$\bar{T}_{sk}$ (area and sensitivity to warming)	$\bar{T}_{sk}$ (area and sensitivity to cooling)
Face	0.07	0.21	0.19
Chest	0.09	0.10	0.08
Upper back	0.09	0.11	0.09
Abdomen	0.18	0.17	0.12
Upper legs	0.16	0.15	0.12
Lower legs	0.16	0.08	0.15
Upper arms	0.13	0.12	0.13
Lower arms	0.12	0.06	0.12
	1.00	1.00	1.00

From Crawshaw et al. (1975).

sweating rate on the thigh and the estimated magnitude of the cold stimulus. The normalized sensitivity coefficients for cold stimuli are shown in Fig. 3.9B. The sequence is somewhat different from that for warm stimuli, but again the forehead is the most sensitive area.

On the basis of these results it is possible to calculate weighting factors for various regions of the body that account for both area and sensitivity coefficients (Crawshaw et al., 1975; Nadel, 1979). By means of these weighting factors (Table 3.II) one obtains a more realistic measure of the thermal drive from the body surface than when measuring average skin temperature.

3.4 Inadequate Stimulation

3.4.1 "PARADOXICAL" SENSATION

On strong heat stimulation one can feel a peculiar quality of cold that has been called "paradoxical" cold sensation by von Frey (1895). This well established phenomenon (Alrutz, 1897; Thunberg, 1901; Hahn, 1927, 1949) is observed when the skin is heated to temperatures above 45 °C.

It is not certain whether a "paradoxical" warm sensation exists (Rein, 1925a; Grundig, 1930).

3.4.2 CHEMICAL STIMULATION

A well-known substance with specific action on thermoreceptors is menthol. It elicits cold sensations at otherwise indifferent skin temperatures. Goldscheider (1898) proved that this cold sensation was not due to evaporative cooling but to a chemical action. Menthol also acts on intravenous appliction (Schwenkenbecher, 1908).

Calcium gives rise to a distinct warm sensation when given intracutaneously or intravenously; it lowers the warm threshold and increases the cold threshold at the same time (Hirschsohn and Maendl, 1922; Schreiner, 1936). Other substances causing warm sensations are locally applied carbon dioxide or carbonic acid (Liljestrand and Magnus, 1922; Gollwitzer-Meyer, 1937) and chloroform, which seems to lower the warm thresholds but also to act on other cutaneous receptors (Rein, 1925a; Schmidt, 1949). Finally, certain substances extracted from spices, such as capsaicine, cinnamonylic-acrylic acid piperidide and undecylenic acid vanillylamide elicit strong warm sensations but also pain (Stary, 1925; Sans, 1949).

4
Cutaneous Thermoreceptors

4.1 Classification of Thermoreceptors

4.1.1 DEFINITION OF RECEPTOR SPECIFICITY

The concept of specific thermoreceptors is based originally on human sensory physiology, in particular on the fact that thermal sensations can be elicited from localized sensory spots in the skin (Blix, 1882). Detailed investigations have revealed a differentiation of "warm" and "cold" spots, that is, local areas responding only with warm or cold sensations. In some cases the spots also responded with thermal sensations to inadequate electrical stimulation. It seems thus justified to speak of specific thermoreceptors in the classical sense of Müller (1840). This means that temperature sensations (E_T) are correlated with localized neural structures (N)

$$E_T \rightarrow N \qquad (12)$$

This type of specificity could be called "phenomenal specificity", since the criterion is a phenomenal quality.

As a result of modern neurophysiological methods it has become possible to define thermoreceptors in physical terms, namely, as nerve endings (N) excited only or preferably by temperature stimuli (S_T)

$$S_T \rightarrow N \qquad (13)$$

Here we can speak of "physical specificity". It is obvious that this concept reaches far beyond the realm of conscious temperature sensation. The physical definition holds for any thermosensitive structure, irrespective of whether its excitation is correlated with temperature sensations, or whether conscious experiences are difficult to establish, as it is the case in animal experiments.

33

Although both concepts are closely related, they are not identical, the criterion being a quality of sensation (E) in the first case and a quality of stimulus (S) in the second case. A clear distinction between both is necessary, especially when the relationship between temperature sensation and neural activity is concerned. For example, a receptor excited by cooling as well as by application of menthol might be classified as a specific cold receptor in terms of sensation (E), but it has at best a relative specificity or selective sensitivity with respect to the thermal stimulus (S). Another example concerns a type of receptor excited by cooling as well as by mechanical stimulation. In the physical sense, such a receptor can be classified as non-specific or bimodal, whereas the corresponding sensation might be a mechanical one in both cases and thus the receptor to be classified as specific or unimodal in the phenomenal sense.

4.1.2 GENERAL PROPERTIES OF THERMORECEPTORS

The first specific thermoreceptors identified by electrophysiological methods were those in the tongue of the cat (Zotterman, 1935, 1936). A quantitative analysis of the discharge of single cold fibres in the cat's tongue (Hensel and Zotterman, 1951b) has revealed some fundamental properties of thermoreceptors that have later been confirmed for many species including man.

Since any biological process is dependent on temperature, and any neural structure in the skin is exposed to thermal stimuli, qualitative data alone are not sufficient to classify a nerve ending as thermoreceptor. In order to establish the physical specificity of a thermosensitive structure, quantitative measurements of the response to various kinds of stimuli are required.

In neurophysiological terms, the general properties of specific cutaneous thermoreceptors can be described as follows: (i) they have a static discharge at constant temperatures (T); (ii) they show a dynamic response to temperature changes (dT/dt), with either a positive temperature coefficient (warm receptors) or a negative coefficient (cold receptors); (iii) they are not excited by mechanical stimuli; (iv) their activity occurs in the non-painful or innocuous temperature range (Hensel et al., 1960; Hensel, 1970, 1973a, 1976a,b).

The variety of cutaneous thermoreceptors can be divided, by the criterion of their dynamic response, into the well-defined classes of warm and cold receptors (Fig. 4.1). Irrespective of the initial temperature, a warm receptor will always respond with an overshoot of its discharge on sudden warming and a transient inhibition on cooling,

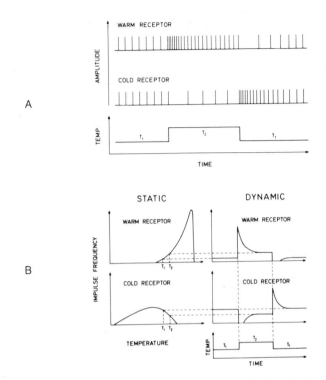

FIG. 4.1. General properties of thermoreceptors. A: Nerve impulses from single warm and cold receptors set up by temperature stimulus. B: Static and dynamic properties of warm and cold receptors as revealed by their response to constant temperatures and temperature changes. (From Hensel, 1974a.)

whereas a cold receptor will respond in the opposite way, namely, with an inhibition on warming and an overshoot on cooling. Besides this dynamic behaviour there are also typical differences in the static frequency curves of both types of cutaneous receptors, in that the temperature of the maximum discharge is much lower for cold receptors than it is for warm receptors.

Table 4.I shows the occurrence of specific thermoreceptors in the skin of various homeotherms. Only those receptors have been included whose specificity has been tested according to the above criteria.

In the skin of human subjects, cold receptors have been identified by electrophysiological methods (Hensel and Boman, 1960). As far as can be concluded from the relative spike height they belong to the group of myelinated fibres. More recently specific warm receptors

TABLE 4.1. Specific thermoreceptors in the external skin of homeotherms.

Species	Nerve	Receptive field	Type of receptor	Author[a]
Man	radial	hairy skin, hand dorsum	cold, warm	Hensel and Boman (1960), Hensel (1976a), Konietzny and Hensel (1975, 1977)
Monkey	median, ulnar saphenous	hairy and glabrous skin, arm and leg	cold (B)	Iggo (1963, 1969), Kenshalo and Gallegos (1967), Perl (1968)
	saphenous, radial	hairy skin, hand and foot dorsum	warm (B)	Hensel (1969), Hensel and Iggo (1971)
Dog	infra-orbital	hairy skin	warm	Sumino et al. (1973)
	infra-orbital	hairy and marginal skin, face	cold (B)	Hensel (1952a), Iriuchijima and Zotterman (1960), Iggo (1969)
			warm	Iriuchijima and Zotterman (1960)
Cat	infra-orbital	hairy and marginal skin	cold (B)	Hensel (1952b), Boman (1958), Iriuchijima and Zotterman (1960)
			warm	Iriuchijima and Zotterman (1960), Hensel (1968), Hensel and Kenshalo (1969)
	saphenous	hairy skin, leg	cold, warm	Hensel et al. (1960)
Rat	infra-orbital	hairy and marginal skin, face	cold	Iriuchijima and Zotterman (1960), Stolwijk and Wexler (1971), Boman (1958)
	saphenous	hairy skin, leg	cold, warm	Iriuchijima and Zotterman (1960)
	scrotal	hairy skin, scrotum	cold (B), warm	Iggo (1969); Hellon et al. (1975)

[a] Only earlier investigators are included.   B: Static burst discharge

have been found by micro-electrode recordings from the human radial nerve (Konietzny and Hensel, 1975, 1977). The conduction velocities of 0.5 to 0.75 m s^{-1} show that these warm fibres are unmyelinated.

4.1.3 UNSPECIFIC RECEPTORS

Various cutaneous receptors respond to temperature but are not to be considered as specific thermoreceptors. Certain slowly adapting (SA) mechanoreceptors (SA I type with dynamic sensitivity, SA II type with static and dynamic sensitivity) served by small myelinated fibres respond to both mechanical stimulation and cooling (Hensel and Zotterman, 1951d; Iggo, 1968, 1969; Chambers et al., 1972; Poulos, 1975). They are also found in humans (Hensel and Boman, 1960; Konietzny and Hensel, 1979).

Receptors excited by intense heating and known variously as "heat receptors" (Iggo, 1959; Beck and Handwerker, 1974; Beck et al., 1974), "thermal nociceptors" (Iggo and Young, 1975; Iggo and Ramsey, 1976), "polymodal nociceptors" (Bessou and Perl, 1969; Burgess and Perl, 1973; Perl, 1976; Beitel and Dubner, 1976; Croze et al., 1976; Kumazawa and Perl, 1977; La Motte and Campbell, 1978), and also described in humans (van Hees and Gybels, 1972; Torebjörk, 1974; Torebjörk and Hallin, 1972, 1974, 1976; van Hees, 1976) are excited by noxious heat above the range of typical warm sensations. Polymodal nociceptors respond also to strong mechanical stimulation and, in some cases, to cooling (La Motte and Campbell, 1978). It might be possible that some of these receptors contribute to the sensory quality of "heat".

4.2 Localization and Structure of Thermoreceptors

4.2.1 PHYSIOLOGICAL MEASUREMENTS OF DEPTH

As long as no morphological evidence was available, one has tried to measure the intracutaneous depth of thermoreceptors by means of physiological methods. These investigations are based on rather vague assumptions and the results thus at best approximative. By use of thermoelectrically controlled intracutaneous temperature movements and measurements of reaction time an average depth of 0.15 to 0.17 mm for cold and 0.3 to 0.6 mm for warm receptors in human subjects was found (Bazett and McGlone, 1930; Bazett et al., 1930). The results suggest that the layer of cold receptors is immediately beneath the epidermis, whereas the site of warm receptors is in the upper layers of the corium.

Electrophysiological measurements of the depth of thermoreceptors are more accurate, as they exclude any error due to central neural processes. When rapid cold jumps with precise start are applied to the skin, and the thermal diffusion coefficient of the tissue is known, one can calculate the velocity of the cold wave penetrating the skin. From the latency of the first cold impulse, amounting to only a few hundredths of a second after the onset of cooling, and the threshold temperature change, a minimum depth of 0.18 ± 0.05 mm for the cold receptors in the cat's tongue was obtained, while the maximum depth was about 0.22 mm (Hensel et al., 1951). This means that the receptors must be situated in a rather superficial subepidermal layer.

4.2.2 MORPHOLOGICAL STRUCTURES

Von Frey's (1895) hypothesis that specific corpuscular nerve terminals were the anatomical substrate of thermal receptors has started numerous attempts to identify histologically the underlying neural structures of cold and warm spots in human subjects. The results of these endeavours were negative (for references see Hensel, 1952a).

Later, cold- and warm-sensitive cutaneous areas in man have been found without any encapsulated or corpuscular nerve endings (Hagen et al., 1953; Weddell et al., 1955; Weddell and Miller, 1962). This rules out von Frey's original concept but does not disprove, of course, the functional and morphological specialization of cold and warm receptors.

Further progress can only be expected by a direct combination of electrophysiological and electron microscopical methods. An attempt in this direction was made in the nasal area of cats (Kenshalo et al., 1971). Further investigations of the ultrastructure of thermoreceptors with an improved technique of intravital fixation were carried out by Hensel et al. (1974). Afferent impulses from single cold and warm fibres dissected from the infraorbital nerve were recorded, and the spot-like receptive fields were localized under the microscope by means of small thermal stimulators with a tip diameter of 0.05 mm. All receptors were carefully tested for specificity. The accuracy of localization was about 0.05 to 0.1 mm. In most cases the cold spots even responded when the stimulator was approached without touching the skin, which proves that the receptor site must be very superficial. After marking the receptive field, the skin was excised and prepared for electron microscopy.

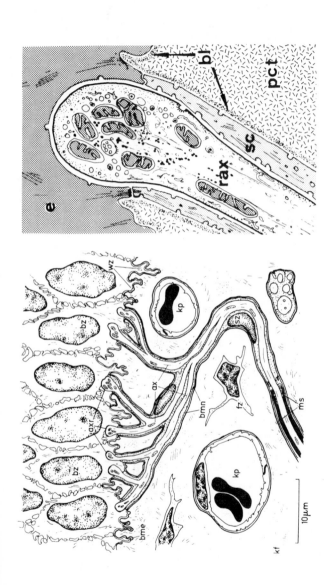

FIG. 4.2. Left: Semischematic diagram of nerve endings at the site of an electrophysiologically identified cold receptor in the hairy skin of the cat nose. ms, Myelin sheath of afferent nerve fibre; sz, non-myelinated Schwann cell accompanying terminal axons (ax) to the basement membrane (bme) of epidermis; bmn, basement membrane of nerve terminals, axr, receptive endings with mitochondria; bz, basal epidermal cells; wz, root feet of basal cells; kp, capillaries; fz, fibrocyte; kf, collagen fibrils of stratum papillare. Right: Schematic representation of a cold receptor axon. The terminal protrudes into a basal epidermal cell. The typical structures of the receptor matrix are present below the receptor membrane. Receptor axon (rax), Schwann cell (sc), epidermis (e), papillary connective tissue (pct), basal lamina (bl). (From Hensel et al., 1974.)

TABLE 4.II

Average conduction velocities of single cold and warm fibres in several species. m, Myelinated fibres: nm, non-myelinated fibres

Species	Nerve	Receptive field	Cold fibres[a] ms^{-1}	Warm fibres[a] ms^{-1}	Author
Man	radial	hairy skin, hand dorsum		0.5–0.8 (2) nm	Konietzny and Hensel (1975)
Monkey	saphenous	hairy skin, leg and foot dorsum	6.3±2.5 (8) m 0.7±0.3 (6) nm	0.7±0.2 (9) nm	Hensel and Iggo (1971)
	radial	hairy skin, hand	8.0±3.0 (16) m		Perl (1968)
	musculo-cutaneous	hairy skin, arm	5.2±1.8 (5) m 0.6 (1) nm		Iggo (1969)
	median	glabrous skin, hand	10.7±3.0 (5) m		Iggo (1969)
	median	glabrous skin, hand	14.5±4.6 (147) m		Darian-Smith et al. (1975)
	median, ulnar	glabrous skin, hand		1.2±0.5 (50) nm	Darian-Smith et al. (1979b)
	median, ulnar saphenous, femoral	glabrous skin, hand, foot		0.8±0.1 (14) nm	Duclaux and Kenshalo (1980)
	infra-orbital	hairy skin, face		2.5 (24) nm 2.5 (16) m	Sumino et al. (1973)

Cat	saphenous	hairy skin, leg	1.0 (3) nm	0.8 (1) nm	Hensel *et al.* (1960)
	post. fem. cut.	hairy skin, leg	0.8–1.1 (8) nm	0.8 (2) nm	Bessou and Perl (1969)
	plantar	glabrous skin, foot pad	5.0–6.8 (3) m	1.1 (1) nm	Burgess and Perl (unpubl.)
		hairy skin, leg	1.0 (1) nm		
	dorsal root ganglion	hairy skin, tail	1.2–1.9 (2) nm	0.8–1.2 (5) nm	
Dog	saphenous	hairy skin, leg		0.4–0.6 (2) nm	Iriuchijima and Zotterman (1960)
	infra-orbital	hairy and marginal skin, face	14 (3) m		Iggo (1969)
Rat	saphenous	hairy skin		1.0±0.2 (7) nm	Iriuchijima and Zotterman (1960)

[a] Number of units indicated in brackets.

In the hairy skin of the cat's nose the receptive structures at the site of cold spots are served by thin myelinated axons dividing into several unmyelinated terminals within the stratum papillare (Fig. 4.2). The terminal axons are accompanied by unmyelinated Schwann cells as far as the epidermal basement membrane. A continuous connection between the basement membrane of the epidermis and that of the nerve terminals is seen. The receptive endings which penetrate a few micrometres deep into the basal epidermal cell contain numerous mitochondria as well as an axoplasmatic matrix with fine filaments and microvesicles. With a certain variability such structures were found regularly at the site of electrophysiologically identified cold receptors in the hairy skin of the cat's nose. In the glabrous skin however, it was difficult to discriminate between the cold-sensitive structures and the axons serving Merkel cells. The anatomical substrate of specific warm receptors has not yet been identified.

4.2.3 AFFERENT INNERVATION AND RECEPTIVE FIELDS

In various species of homeotherms specific cold receptors are innervated by fairly thin myelinated fibres (Aδ group according to the classification by Erlanger and Gasser (1937), or group III according to Lloyd and Chang (1948) as well as by unmyelinated fibres (C fibres, or group IV fibres, respectively). On the basis of investigations in monkeys (Hensel and Iggo, 1971) and available evidence in man (Hensel and Boman, 1960) it seems possible that human cold receptors are served both by thin myelinated and unmyelinated fibres. Specific warm fibres in humans were found to be unmyelinated (Konietzny and Hensel, 1975). Table 4.II summarizes some findings on conduction velocities of single cold and warm fibres in the skin of several homeothermic species. We can conclude that cutaneous cold fibres with the highest conduction velocities of 20 m s^{-1} (Iggo, 1959) have diameters of 3 to 4 μm (Maruhashi et al., 1952), while the diameters of the slowly conducting C fibres are in the range of 1 μm.

In most cases single cold and warm fibres, respectively, innervate one peripheral spot in the skin smaller than 1 mm^2 (Iggo, 1969; Hensel and Kenshalo, 1969; Hensel, 1969, 1973a; Sumino et al., 1973; La Motte and Campbell, 1978; Darian-Smith et al., 1979b). In the cat's nose the cold-sensitive spots amounted in some cases to a few tenths of a millimetre and had a spatially inhomogenous cold sensitivity. Only in one investigation primate single cold fibres were reported to innervate up to eight multiple spots, the whole field amounting to 1.7 cm^2 (Kenshalo and Gallegos, 1967). Spot-like receptive fields were also found for single warm fibres in humans (Konietzny and Hensel, 1975).

4.3 Thermoreceptor Excitation and Temperature

4.3.1 STATIC AND DYNAMIC ACTIVITY OF COLD RECEPTORS

At constant skin temperature in the normal range all cutaneous cold receptors exhibit a static discharge with constant impulse frequency

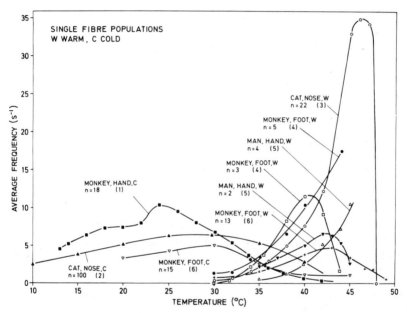

FIG. 4.3. Average static discharge frequency of populations of cold and warm fibres as function of skin temperature. ((1) From Dykes, 1975; (2) from Hensel and Schäfer, unpublished; (3) from Hensel and Kenshalo, 1969; (4) from Hensel, 1973a; (5) from Konietzny and Hensel, 1979; (6) from Kenshalo, 1976.)

(Hensel and Zotterman, 1951a). The temporal sequence of impulses can be more or less regular, or it can consist of periodic bursts of 2 to 12 impulses separated by silent intervals. The temperature range of static activity varies for different cold fibres, the extreme skin temperatures being −5 and 43 °C. At these extremes, of course, the temperature of the receptor must not necessarily be the same as that of the skin surface. The static impulse frequency of individual cold receptors rises with temperature, reaches a maximum and falls again

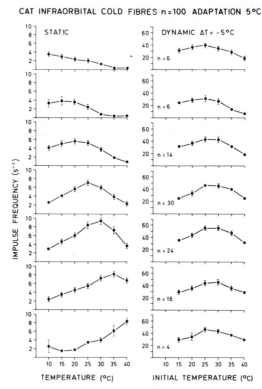

CAT INFRAORBITAL COLD FIBRES n=100 ADAPTATION 5°C

FIG. 4.4. Average static discharge frequency and maximal dynamic response of single cold fibre populations from the cat nose as function of temperature. Dynamic responses were obtained by cooling steps of 5 °C. Groups according to static maxima at 10, 15, 20, 25, 30, 35 and 40 °C, n = number of fibres in each group. Cats were adapted to ambients of 30 °C for 4 years. Bars indicate standard error of mean. (From Hensel and Schäfer, unpublished.)

at high temperatures. For various cold fibres in different species the static maxima are scattered over a temperature range from −5 to 40 °C. Thus a whole receptor population will cover a broader span of temperature than a single receptor. The average static maxima of larger cold fibre populations in various skin areas are rather similar, ranging from 25 to 30 °C in monkeys, cats and rats (Fig. 4.3). Only in the lip of dogs the average value was near 35 °C (Iggo, 1969; Iggo and Iggo, 1971). As to human cold fibres, there is not enough information available for plotting an average curve of the static discharge.

The majority of specific cold fibres in the nose of cats had static maxima at 25 to 30 °C, but there were also cold fibres with maxima at 20, 15, and 10 °C, and a few fibres had maxima at 40 °C (Fig. 4.4).

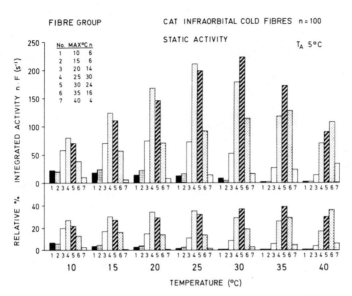

FIG. 4.5. Across-fibre pattern of static activity for seven groups of cold fibres from the cat nose at various constant temperatures. Groups are indicated in the inset. Upper graph: integrated activity (average frequency multiplied with number of fibres for each group). Lower graph: relative integrated activity in percent. Cats were adapted to ambients of 5 °C for 4 years. (From Hensel and Schäfer, unpublished.)

From these data an across-fibre pattern is drawn for various constant temperatures (Fig. 4.5). For seven groups classified by their static maxima between 10 and 40 °C, the integrated frequency, that is, the average frequency multiplied with the number of fibres in each group, is depicted. Obviously there is a considerable change of the across fibre pattern with temperature, in particular in the lower range.

In the glabrous skin of the cat's nose there are cold-sensitive fibres with even lower static temperature ranges (Duclaux et al., 1980). A number of fibres had static maxima at 5 °C and even at −5 °C. When cooling steps of 5 degrees were applied, the fibres with the lowest static maximum showed a maximal dynamic response to cooling from 0 to −5 °C. None of the receptors did respond to moderate mechanical stimulation.

On sudden cooling to a lower temperature level, the cold receptors respond with a transient overshoot in frequency, followed by partial adaptation to the new static discharge (Figs 4.6 and 4.7). When the skin is rewarmed to the initial level, a transient decrease in frequency or silent period is seen, after which the frequency rises again and finally reaches the initial static value. The highest dynamic frequency

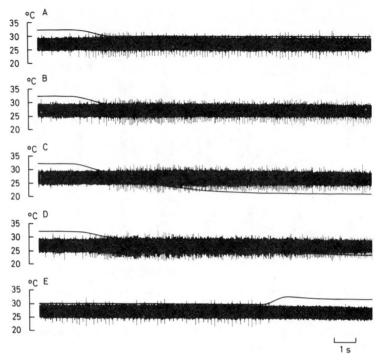

Fig. 4.6. Afferent impulses of a single cold fibre from human hairy skin and skin temperature when applying various cold stimuli. (From Konietzny and Hensel, unpublished.)

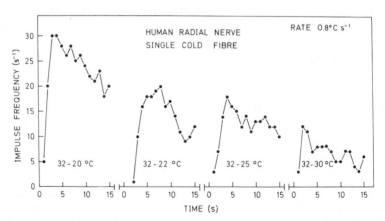

Fig. 4.7. Discharge frequency of a single cold fibre from human hairy skin when applying linear temperature decreases of 0.8 °C s^{-1} and various magnitude. (From Konietzny and Hensel, unpublished.)

observed in a single cutaneous cold fibre was 300 s^{-1} (Iggo, 1969), the highest ratio of static to dynamic frequencies 1:30.

When equal cooling steps are applied at various adapting temperatures, the dynamic overshoot is a function of temperature and follows approximately the shape of the static activity curve (Hensel and Zotterman, 1951b; Iggo, 1969; Darian-Smith and Dykes, 1971; Kenshalo et al., 1971; Kenshalo, 1976) (Fig. 4.8).

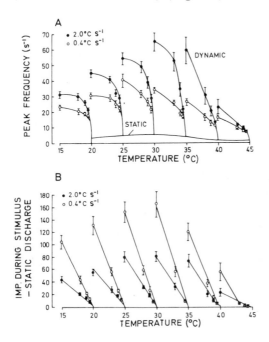

Fig. 4.8. A: Mean peak frequencies of responses of 15 cold units in monkeys to cooling of 0.5, 1.0, 2.0, and 5.0 °C as a function of adapting temperature. The parameter is rate of cooling. B: Means of the total impulses during stimulation; the parameters are the same as in A. Brackets indicate standard error of mean. (From Kenshalo and Duclaux, 1977.)

The higher the rate of cooling at a given adapting temperature, the higher is the dynamic response (Hensel, 1973a; Molinari and Kenshalo, 1977). Likewise, the dynamic response of cold units increases with the magnitude of a cold stimulus, the relationship between stimulus and response magnitude depending on species, index of dynamic response and shape of stimulus. Using rapid cold steps, the peak frequency of the dynamic overshoot (Hensel, 1953a) or the number of spikes during a 4 s rectangular cooling step (Darian-Smith and

Dykes, 1971) was found to be a linear function of the stimulus magnitude within a range of 5 °C, sometimes 10 °C. When slower linear changes of 0.4 to 2 °C s^{-1} followed by a plateau were applied to the skin of monkeys, the peak frequencies increased in a non-linear way with the stimulus magnitude (Fig. 4.8). However, if the total number of impulses during the cooling phase was used as criterion, a linear relationship was found (Kenshalo and Duclaux, 1977).

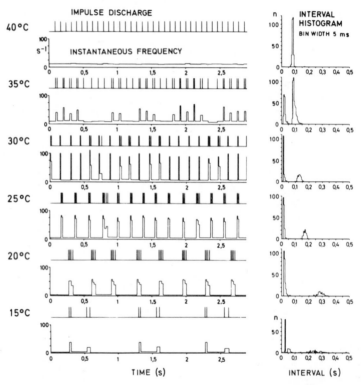

Fig. 4.9. Impulse discharge, instantaneous frequency and interval histograms of a single cold fibre from the cat lingual nerve at various constant temperatures. (From Bade *et al.*, 1979.)

From the slope of the static and dynamic frequency curves we can derive the static and dynamic differential sensitivity ($\Delta F/\Delta T$), i.e. the change in frequency (ΔF) for a small change in temperature (ΔT). Of course, the dynamic differential sensitivity increases with the rate of change. For the cold receptors in the cat's nose, a maximum static differential sensitivity of -1 s^{-1} °C^{-1} and a maximum dynamic differential sensitivity of -50 s^{-1} °C^{-1} was found (Hensel and Kenshalo, 1969). For the monkey's skin, the average differential sensitivity of a cold fibre population amounted to -21 s^{-1} °C^{-1} (Kenshalo, 1976).

4.3.2 BURST DISCHARGE

At certain temperatures, many cold receptors show a temporal sequence of impulses consisting of periodic bursts of 2 to 12 impulses interrupted by pauses (Fig. 4.9). The burst discharge has first been described for cold fibres in the lingual nerve of cats (Hensel and Zotterman, 1951b; Dodt, 1953) and later for cold receptors in various species including man (Table 4.III). While in the skin of monkeys (Iggo, 1969; Hensel and Iggo, 1971) and in the tongue of cats (Bade et al. 1979) nearly all cold receptors show static bursting, this phenomenon is observed only in about 25% of the cat's nasal cold fibres, but most of the non-bursting fibres respond with transient bursting during dynamic cooling (Hensel and Schäfer, 1979). In human subjects only some occasional observations of bursting cold fibres during dynamic cooling have been made (Hensel, 1976a).

Most of the burst parameters, such as burst frequency or number of impulses per burst, increase monotonically with decreasing constant temperature in the bursting range (Iggo, 1969; Poulos, 1971, 1975; Iggo and Young, 1975; Dykes, 1975; Dubner et al., 1975; Bade et al., 1979). The burst discharge usually starts at the temperature of the highest static average frequency and continues to temperatures as low as 20 °C, sometimes even 10 °C. An example for the change of various burst parameters with temperature of a population of cold fibres in the tongue of the cat is shown in Fig. 4.10.

At low temperatures, the burst discharge becomes irregular, the intervals being more or less randomly distributed (Fig. 4.11). With increasing temperatures, regular bursts occur that are gradually transformed into a regular sequence of single impulses (Fig. 4.9). Finally the discharge becomes irregular again; however, the longer intervals are not randomly distributed but are multiples of the shorter ones, as can be seen in the interval histograms (Fig. 4.11).

Certain cold receptors, preferably in the nose of the cat (Hensel, 1973a; Kenshalo, 1975), show a regular discharge at constant temperatures but respond with a transient burst discharge during rapid cooling (Fig. 4.19). The sequence of bursts has the appearance of a damped oscillation (Braun et al., 1978). Similar dynamic bursts have also been observed in human cold fibres (Hensel, 1976a) and in cold

TABLE 4.III

Burst parameters of cold fibres in various species

Author	Nerve	Spikes/burst (SB)	Burst duration (BD)	Burst period (BP)	Range of regular burst pattern °C
		n	ms	ms	
Iggo (1969)[a]	Monkey, N. medianus	2–4		285–650	31–21
Poulos (1971)[a]	Monkey, N. trigeminus	2–5	20–100	130–340	33–21
Dykes (1975)	Monkey, N. medianus	2–4	12–220	240–600	26–20
Hellon et al. (1975)	Rat, N. pudendus	2–4			27–20
Bade et al. (1979)	Cat, N. lingualis	2–4	15–100	160–520	30–15

[a] Partially extrapolated data.

fibres in the monkey (Kenshalo, 1976), the latter showing also a static
burst discharge.

4.3.3 STATIC AND DYNAMIC ACTIVITY OF WARM RECEPTORS

At constant temperatures, warm receptors exhibit a static discharge
that begins in the range above 30 °C, increases its frequency with
rising temperature and decreases again at still higher levels (Fig. 4.3).

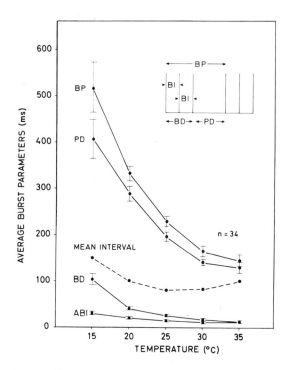

FIG. 4.10. Average burst parameters and mean spike interval from the cat lingual
nerve as function of constant temperature. BD, Burst duration; PD, pause duration;
BP, burst period; BI, intraburst interval; ABI, average intraburst interval; n, number
of cold fibres. Bars indicate standard error of mean. (From Bade et al., 1979.)

The maximum static frequency of individual warm receptors in
various species is scattered over a temperature range from 41 to 47 °C.
For warm receptors in the skin of monkeys, the static maximum
frequency is considerably higher than that for cold receptors. As can
be seen in Fig. 4.3, there is some overlap of the static discharges of
warm and cold receptor populations.

The average maximum frequency of warm receptors in the cat's nose (Hensel and Kenshalo, 1969), certain warm receptors in the monkey's skin (Hensel, 1969, 1973a; Sumino *et al.*, 1973; Kenshalo, 1976) and in the human hand (Konietzny and Hensel, 1975, 1979) is in the range between 45 and 47 °C. Other populations of warm fibres in the scrotum of the rat (Iggo, 1969; Hellon *et al.*, 1975) and in the skin of monkeys (Hensel, 1973a; Sumino *et al.*, 1973) and humans

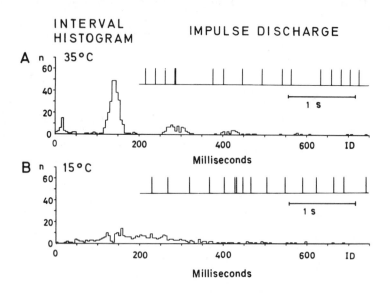

Fig. 4.11. Afferent impulses and interval histograms of a bursting cold fibre from the cat lingual nerve at various constant temperatures. A: Non-bursting range at high temperature; B: non-bursting range at low temperature. (From Braun *et al.*, 1980.)

(Fig. 4.12) have their average static maxima in the temperature range between 41 and 43 °C. For a larger population of warm fibres in the monkey's skin, the average static maximum was at 43 °C (Kenshalo, 1976; Duclaux and Kenshalo, 1980). The highest static differential sensitivity $(\Delta F/\Delta T)$ of human warm fibres was about $+4\ \mathrm{s}^{-1}\ {}^{\circ}\mathrm{C}^{-1}$ (Konietzny and Hensel, 1975). In contrast to the cold fibres, warm fibres show a fairly regular sequence of impulses in the steady state, and only in the high temperature range, when the activity starts to decrease, the discharge may become more irregular.

On sudden temperature changes, warm receptors behave in the opposite way than do cold receptors, in that they respond to warming with an overshoot and to cooling with a transient decrease of the discharge frequency. The highest dynamic frequency of a single warm fibre from the cat's nose was 200 s⁻¹, that is 5.5 times higher than the average static maximum of 36 s⁻¹. The dynamic differential sensitivity amounted to $+70 \text{ s}^{-1} \text{ }^{\circ}\text{C}^{-1}$ (Hensel and Kenshalo, 1969).

A marked dynamic overshoot on step-like warming was found for warm fibres in cats (Hensel and Huopaniemi, 1969), monkeys

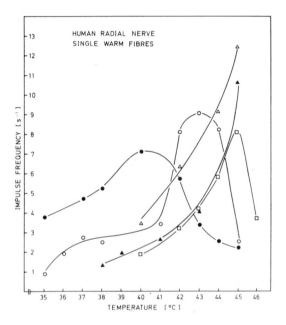

Fig. 4.12. Static discharge frequency of single warm fibres from human hairy skin of the hand as function of constant temperature. (From Konietzny and Hensel, 1979.)

(Sumino et al., 1973; Kenshalo, 1976; Darian-Smith et al., 1979b; Duclaux and Kenshalo, 1980), and humans (Konietzny and Hensel, 1977, 1979). In the palmar skin of monkeys, the total number of impulses during 4 s after the step increase was a linear function of stimulus magnitude (Darian-Smith et al., 1979b). A similar relationship was found for warm fibres in the monkey's face (Sumino et al., 1973).

When the temperature is linearly increased with different slopes from equal adapting levels to equal higher levels, the transient overshoot in frequency of human warm receptors increases with the rate of change, the instantaneous frequency during the rising phase of the stimulus being an almost linear function of temperature (Fig. 4.13). Starting from one and the same adapting temperature with a constant rate of warming, the magnitude of the dynamic response of these receptors increases with the magnitude of the temperature increment (ΔT) in an approximately linear correlation (Fig. 4.14). The slope of this curve is also a function of adapting temperature: the higher the adapting level, the larger the dynamic response as a function of temperature increment. These properties of human warm

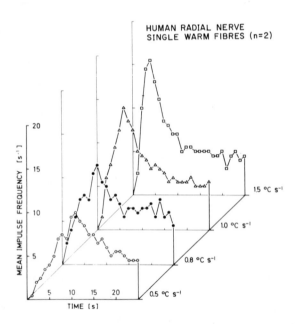

Fig. 4.13. Average impulse frequencies of two single warm units from human hairy skin at various rates of warming (dT/dt = 1.5, 1.0, 0.8 and 0.5 °C s^{-1}) from 32 to 37 °C. (From Konietzny and Hensel, 1977.)

fibres (Konietzny and Hensel, 1977, 1979) closely resemble those recently described for warm fibres in the palm of monkeys (Darian-Smith *et al.*, 1979b). Similar results were also reported for warm fibres in the monkey's face (Beitel *et al.*, 1977).

For warm fibres in the monkey's skin, the number of impulses during a linear increase in temperature was found to be a linear function of stimulus magnitude at adapting temperatures around 40 °C (Fig. 4.15), while at any other condition the relation was non-linear (Kenshalo, 1976; Duclaux and Kenshalo, 1980).

A particular type of response was found for a group warm fibres in the monkey's face having static maxima around 45 °C. Rapid steps to noxious temperatures above 45 °C elicited short-latency, high-frequency initial transient responses which were followed by suppression of activity. These sudden increases in activity are thought to

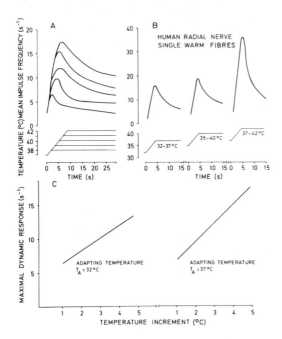

FIG. 4.14. A: Impulse frequency of a single warm fibre from human hairy skin when warming with constant rate ($dT/dt = 0.8$ °C s^{-1}) and various magnitude ($\Delta T = 1$, 2, 3, 4, and 5 °C). Adapting temperature 37 °C. B: Impulse frequency of a single warm fibre and skin temperature when warming with constant rate ($dT/dt = 1.5$ °C s^{-1}) and equal magnitude ($\Delta T = +5$ °C). Adapting temperature 32, 35 and 37 °C. C: Maximal dynamic response of the same unit as shown in A. Adapting temperature 32 and 37 °C. (From Konietzny and Hensel, 1977.)

represent the peripheral neural event associated with the stinging, brief quality of "first pain" (Sumino *et al.*, 1973).

4.3.4 SPATIAL TEMPERATURE GRADIENTS

From the neurophysiological data we can conclude that the excitation of cutaneous thermoreceptors depends (i) on the absolute temperature (T) and (ii) on the rate of temperature change (dT/dt) or the temporal gradient. The hypothesis of spatial temperature gradients (dT/dx) being the adequate stimulus of thermosensitive nerve endings could not be verified by cooling the tongue of the cat with reversed gradients (Hensel and Zotterman, 1951c; Hensel and Witt, 1959). When a cold receptor on the upper surface of the tongue was cooled from above, its discharge frequency increased. On cooling the tongue from the lower surface, however, the resting discharge of the cold receptor increased as well, in spite of a reversed spatial temperature gradient. In accordance with sensory experiments (p. 25) it must be concluded that the adequate stimulus of thermoreceptors is temperature and its temporal change *per se* and not the direction or slope of a spatial gradient.

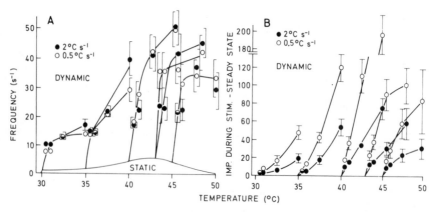

FIG. 4.15. Two indexes of response magnitude to warming episodes from each of the five adapting temperatures in monkeys. A: Mean peak frequencies of responses of 15 warm units to warm stimuli. Starting at the 30 °C adapting temperature (AT), for example, reading from left to right, the first vertical pair of points (one open, one filled) represents the mean peak frequency of response to +0.5 °C at the slow and the fast rate from the 30 °C AT. The second vertical pair represents the mean responses to +1.0 °C warming: the third, the mean response to +2.5 °C: while the fourth, the mean response to +5.0 °C. The sequence is repeated to each AT. The parameter is rate of warming. Brackets indicate the standard errors. B: The same as A except that the index used here is total impulses during stimulation (duration of the temperature change). (From Duclaux and Kenshalo, 1980.)

4.4 Inadequate Stimulation

4.4.1 PARADOXICAL DISCHARGE

The occurrence of "paradoxical" excitations was reported for cold fibres from the lingual nerve of the cat (Dodt and Zotterman, 1952).

Whereas at temperatures between 40 and 45 °C no static discharge of these fibres is seen, they start again discharging at lingual temperatures above 45 °C. On further warming, the frequency of this paradoxical excitation rises and reaches a maximum at about 50 °C. Above this value the receptors will be damaged. For warm receptors no static paradoxical discharge has been found at low temperatures.

4.4.2 CHEMICAL STIMULATION

Menthol causes a shift of the activity range of lingual cold receptors towards higher temperatures. When the cat tongue is kept at 40 °C there is no static activity in most cases. Application of menthol dissolved in water ($1:10^4$) elicits a strong discharge of cold impulses that can be abolished by further warming (Hensel and Zotterman, 1951e). Menthol enhances not only the normal activity of cold receptors but also the paradoxical discharge at 45 to 50 °C (Dodt et al., 1953).

Acetylcholine in small doses extends the static activity range of cold fibres in the lingual nerve of the cat towards higher temperatures, whereas higher doses inhibit the static discharge (Dodt et al., 1953). Increased concentrations of carbon dioxide reduce the discharge frequency of lingual cold receptors and stimulate at the same time receptors sensitive to warming (Dodt, 1956; Boman et al., 1957). This corresponds well with the subjective experience in a CO_2 bath.

The effects of ouabaine, calcium and EDTA will be discussed in connection with the transducer mechanisms of thermoreceptors.

When the blood supply to the tongue is arrested, the steady discharge of cold receptors in the tongue gradually diminishes and often falls to zero within 2 to 4 min (Hensel, 1953b). This corresponds well with the observation in humans that ischaemia will lead to a decrease of cold sensation (Ebbecke, 1917; Issing et al., 1978). Release of the arterial occlusion is followed by a transient overshoot and complete restoration of the initial discharge within 1 min. This behaviour of cold receptors is in accordance with the subjective experience known as "Ebbecke's phenomenon" (Ebbecke, 1917): When an extremity is cooled and the blood flow arrested for several minutes, the following release of the occlusion is accompanied by an intense cold sensation.

4.5 Transducer Mechanisms of Thermoreceptors

It is obvious that thermoreceptors show a continuous discharge even when the temporal and spatial temperature gradient is zero, that is, no thermal energy is transferred between the environment and the receptor. Of course, no static discharge is possible without energy but this energy may be derived from biochemical processes within the receptor rather than from the transfer of thermal energy, the temperature being only a controlling factor for metabolic rates and other events.

Until recently the question how temperatures and temperature changes are transformed into neural activity in a thermoreceptor was a matter of mere speculation. In order to describe the bell-shaped frequency/temperature curve of the static discharge as well as the transient overshoot and undershoot in frequency on dynamic temperature changes, more than one temperature-dependent process must be assumed.

Sand (1938) proposed a model for the response of a cold-sensitive receptor. He postulated two processes with positive temperature coefficients, one process being excitatory, the other inhibitory. The actual response corresponds to the difference between both processes. On sudden cooling, the inhibitory process rapidly falls to a lower level, while the excitatory process reaches a lower level more slowly. Thus a dynamic overshoot in frequency and a decline to a new static level will result. The opposite occurs during rapid rewarming, leading to a transient decrease in frequency with subsequent rise to the initial static level. This model was purely formal in nature, and so were similar models for thermoreceptors proposed by others (for references see Hensel, 1973a).

Observations of temperature-dependent discharges of bursting and non-bursting pacemaker neurons in molluscs, such as Aplysia and Helix (for references see Braun et al., 1980), in combination with recordings of cold receptor discharges under ouabain and potassium-free extracellular solutions led to the hypothesis that the inhibitory process of the cold receptor may be the activity of an electrogenic sodium pump with a positive temperature coefficient and thus leading to hyperpolarization of the receptor membrane with increasing temperature. The excitatory process was assumed to be the Na^+/K^+ permeability ratio which increases also with temperature and thus leads to increasing depolarization of the receptor membrane. The static and dynamic impulse frequencies were assumed to depend on the difference between the two processes (Pierau et al., 1975). Blocking the sodium pump by ouabain leads to an increase in the static cold receptor discharge, a diminished dynamic overshoot on cooling and an increased frequency on dynamic warming (Pierau et al., 1975). This agrees well with the assumption that the sodium pump corresponds to the inhibitory process of Sand's model.

By analogy from intracellular recordings of bursting neurons in molluscs we may assume that bursts in a cold receptor are triggered by an oscillating receptor potential exceeding a threshold of depolarization. The frequency of bursts will depend on the frequency of the oscillation, while the intraburst frequency will increase with the amount of depolarization during each phase of the oscillation. At a given frequency of the oscillation, an increasing depolarization will also lead to a higher number of spikes per burst.

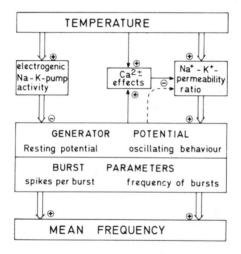

FIG. 4.16. Hypothetical transducer mechanisms of bursting cold receptors. For further explanation see text. (From Braun *et al.*, 1980.)

Figure 4.16 shows some hypothetical receptor mechanisms (Bade *et al.*, 1978; Braun *et al.*, 1980). The resting potential may depend on the activity of an electrogenic $Na^+ K^+$ pump, the oscillations on periodic changes of the Na^+/K^+ permeability ratio. The sodium-potassium permeability ratio is assumed to be additionally affected by a voltage- and temperature-dependent effect of calcium on the potassium permeability. It is further assumed that a non-linearity of the calcium-potassium system related to the membrane potential exists, and that this may act as a negative feedback system: depolarization activates the calcium-potassium system which, in turn, causes repolarization.

From the burst pattern at various temperatures, particularly in the higher temperature range (cf. Fig. 4.11), we can conclude that the oscillation of the receptor potential is maintained throughout, and even beyond, the whole range of static activity, its frequency and amplitude increasing with rising temperature. At higher temperatures, however, an increasing hyperpolarization of the resting potential occurs, due to the activity of a temperature-dependent electrogenic $Na^+ K^+$ pump, so that the peaks of the oscillation finally fail to reach the threshold for the initiation of impulses (Fig. 4.17).

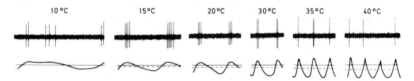

FIG. 4.17. Hypothetical mechanisms of temperature transduction of bursting cold receptors. Further explanation see text. (From Braun *et al.*, 1980.)

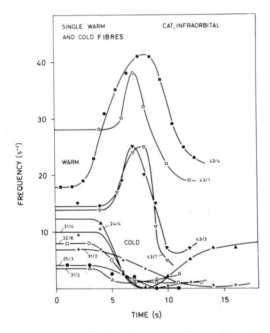

FIG. 4.18. Discharge frequency of single warm fibres (upper four curves) and cold fibres (lower six curves) from the cat nose during intravenous injection of calcium at constant temperatures of the receptive field. The two numbers at each curve indicate the skin temperature (°C) and the dose of Ca^{2+} (mg kg^{-1}), respectively. The injection of calcium (duration 5 s) started about 8 s before the onset of the frequency change. (From Hensel and Schäfer, 1974.)

4.5.1 EFFECT OF CALCIUM AND EDTA

Intravenous injection of calcium inhibits the activity of cold receptors and enhances that of warm receptors (Hensel and Schäfer, 1974) (Fig. 4.18). Intravenous infusion of calcium in cats (ca. 0.3 mg Ca^{2+} min^{-1} kg^{-1}) leads to a considerable decrease in the static

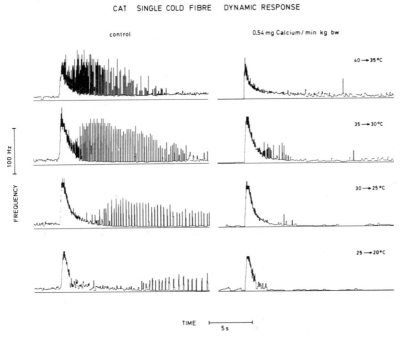

CAT SINGLE COLD FIBRE DYNAMIC RESPONSE

FIG. 4.19. Instantaneous frequency of a single cold fibre from the cat nose during rapid cooling of the skin. Left diagrams: Controls. Right diagrams: Intravenous infusion of Ca^{2+}. (From Schäfer, 1980.)

discharge frequency of cold fibres and a shift of the maximum to lower temperatures (Schäfer et al., 1978). Lowering the calcium level by i.v. infusion of the chelating agent EDTA (bisodium salt of ethylene diamine tetraacetic acid, ca. 3 mg min^{-1} kg^{-1}) causes a marked increase in the static frequency, the effect being greater at higher temperatures. In cold fibres with a dynamic burst discharge on rapid cooling, only the bursts are abolished (Fig. 4.19), whereas the non-bursting initial overshoot remains unchanged (Schäfer et al., 1978).

   It was found that calcium also inhibits static bursting of lingual cold fibres in cats, while EDTA increases the number of spikes per burst (Pierau *et al.*, 1977). Moreover, infusion of Ca^{2+} can transform the burst discharge of a cold receptor into a regular sequence of impulses, the intervals of which correspond to the burst periods (Fig. 4.20). When the calcium level is lowered by i.v. infusion of EDTA, the opposite occurs: a non-bursting cold fibre discharge can be transformed into a bursting one (Schäfer *et al.*, 1979), the intervals of the

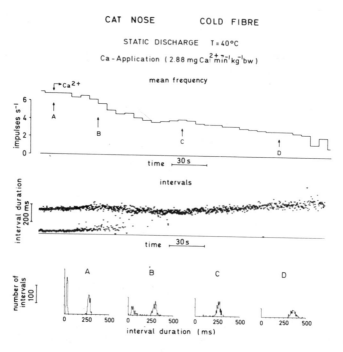

FIG. 4.20. Static discharge of a single cold fibre from the cat nose at 40 °C before (A) and after (B, C, D) i.v. infusion of Ca^{2+}. Upper diagram: Mean discharge frequency as function of time. Middle diagram: Interval duration as function of time. Lower diagram: Interval histograms at various times A, B, C, D. (From Schäfer, 1980.)

regular discharge again corresponding to the burst period (Fig. 4.21). If the burst parameters, such as frequency of bursts, impulses per burst, and intraburst frequency, are plotted as a function of constant temperature there is no essential difference between naturally bursting cold fibres in the cat's tongue and cold fibres in the cat's nose with artificial induction of bursting by EDTA (Schäfer *et al.*, 1979).

This suggests that the underlying processes are the same and that calcium plays an important role in the generation of the burst discharge of cold receptors.

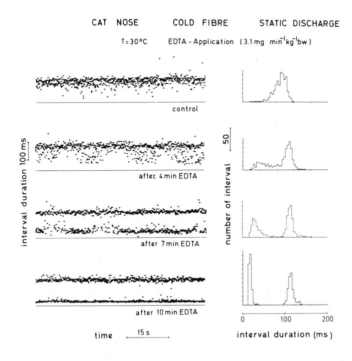

FIG. 4.21. Static discharge of a single cold fibre from the cat nose at 30 °C. Left: Interval duration as function of time. Right: Interval histograms. Uppermost diagrams: Controls. Following diagrams: Four, 7 and 10 min after i.v. infusion of EDTA. (From Schäfer, 1980.)

5
Thermal Afferent Pathways

Our knowledge of the central pathways from peripheral thermoreceptors is rather fragmentary. Evidence in this field is derived from anatomical, histological, histochemical, fluorescence microscopic and autoradiographic investigations, from electrophysiological recordings of neural activity, and from the results of damage to central nervous structures, either by artificial lesions and surgery, or by accidents and neural diseases.

The processing of thermal afferents in higher-order neurons is mainly assessed by single-unit recordings with microelectrodes. Here we are confronted with difficulties and complications that do not exist for peripheral thermoreceptors: (1) Most of the central neurons responding to peripheral temperatures show a high degree of convergence from thermoreceptors as well as from thermally sensitive mechanoreceptors and nociceptors (for references see Burton, 1975; Iggo and Ramsey, 1976; Zimmermann, 1977). Therefore, quantitative relationships between thermal stimuli and neural responses are difficult to establish. When thermoreceptive afferents from a large area of skin converge on a single neuron, it can be difficult to apply homogenous thermal stimuli over the whole area of convergence. (2) Central neurons are more influenced by descending activity (Brown et al., 1973; Tapper et al., 1973; Burton, 1975; Zimmermann, 1977; Carstens et al., 1979) and by anaesthesia than are peripheral thermoreceptors, and any procedure to abolish descending influences might alter the properties of the neuron in question. (3) Except for some recordings of evoked potentials from the human scalp, our knowledge of central nervous activity set up by thermal stimulation is derived from animal experiments, rendering the situation in humans a matter of uncertain analogies.

5.1 Spinal Cord

In higher mammals, primary thermosensitive fibres of the myelinated (Aδ) and the unmyelinated (C) group enter the spinal cord via the spinal ganglion and the dorsal root. In the corresponding segment the afferent fibres are connected to second-order neurons in the dorsal horn (for references on fibre terminations see Kumazawa *et al.*, 1975; Light and Perl, 1977, 1979a,b; Light *et al.*, 1979; Réthelyi *et al.*, 1979), preferably in the more superficial layers (Christensen and Perl, 1970; Hellon and Mitchell, 1974; Iggo and Ramsey, 1976). From the dorsal horn, spinal thermosensitive afferents are mainly conveyed in the contralateral anterolateral (spinothalamic) tract, but there may also be projections via the ipsilateral dorsolateral (spinocervical) tract (Brown, 1973; Iggo and Ramsey, 1976; Cervero *et al.*, 1977). Spinal as well as trigeminal thermoreceptive pathways project to third-order neurons in the ventrobasal complex of the thalamus (Landgren, 1970; Hellon and Misra, 1973b; Poulos, 1975).

Among dorsal horn neurons that are not excited by innocuous temperatures of the skin, units have been found that respond only to mechanical stimulation (class 1), to mechanical stimulation and noxious heat (class 2) and to noxious stimulation (class 3) (for references see Handwerker *et al.*, 1975; Boivie and Perl, 1975; Iggo and Ramsey, 1976).

Specific cutaneous thermoreceptors send small myelinated (Aδ) and non-myelinated (C) afferents into the spinal cord via the spinal nerves. In Rexed's (1952) lamina I of the lumbar dorsal horn in monkeys, "cold" neurons were found that responded to moderate cooling of their receptive fields in the foot and the lower leg (Iggo and Ramsey, 1976). These cold neurons could not be excited by light mechanical stimulation, nor were they excited by skin temperatures in the noxious range. Some of the cold units were found to project into the contralateral anterolateral (spinothalamic) tract.

A characteristic response of a dorsal horn cold neuron is shown in Fig. 5.1. When the centre of the receptive field was rapidly cooled by a thermode of 1 cm in diameter, there was only a weak dynamic response and very little adaptation. However, the weak dynamic response may be due to the small dimension of the thermode, and the slow adaptation to the diffusion of cold around the thermode, thus causing recruitment of cold receptors converging on the same neuron.

As can be seen in interspike histograms, the discharge of these cold neurons is fairly irregular and bears no relation to the typical burst pattern of primary afferent cold units in monkeys (Iggo and Ramsey,

1976). This finding can easily be explained by convergence. The possible number of converging peripheral cold units within the receptive field of a dorsal horn cold neuron may be as many as 100, and this level of convergence will affect the possibility of any preservation of afferent impulse patterns.

These results were confirmed by Kumazawa and Perl (1978), who found in the marginal zone of the dorsal horn in rhesus monkeys a number of units that responded specifically to cooling the skin.

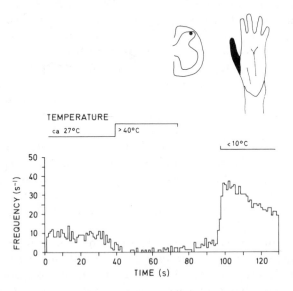

FIG. 5.1. Responses of a dorsal horn cold neuron in a vervet monkey at different stimulus temperatures of the skin. The inset diagrams show the location of the neuron in lamina I of the lumbar spinal cord and the receptive field (black) on the great toe of the foot. The discharge of the neuron is indicated in histograms of 1 s, and the stimulus (a moistened cotton swab) temperatures are shown as bars above the histograms. (From Iggo and Ramsey, 1976.)

Moderate mechanical stimulation was ineffective but in some cases the background discharge of these units was inhibited by gentle mechanical stimulation of skin areas outside the thermal receptive field. As was concluded from the results of electrical stimulation of the dorsal roots, the cold units receive afferents from small myelinated (Aδ) fibres.

Dorsal horn neurons responding to moderate cooling of the receptive field but also to mechanical stimulation or to noxious heat were found in cats and squirrel monkeys (Christensen and Perl, 1970;

Burton, 1975). In cats the location was in lamina I and III–VIII, in squirrel monkeys in I and III–VI (Burton, 1975). It cannot be decided to what extent these neurons are driven by afferents from cold-sensitive slowly adapting I and II mechanoreceptors, and, therefore, it is questionable whether their activity reflects the transmission of afferents from specific cutaneous cold receptors. Single neurons responded to cooling steps with a rapid dynamic overshoot in frequency, in some cases 15 times higher than the static value. The subsequent adaptation had half-times between 2.5 and 5 s.

Another group of dorsal horn neurons responded to innocuous warming as well as to mechanical stimulation of the skin, and a high percentage of these neurons responded to both warming and cooling (Burton, 1975). When a warming step was applied to the skin, peak firing rates were attained within several seconds, and these responses tended to persist. The activity of these neurons was ascribed to a population of peripheral receptors that do not show maximal static activity during innocuous warm stimuli.

Pathways to the cortex from warm and cold receptors in the rat's scrotum will not be discussed here in detail, because they show some particular properties and are evidently not a good model for describing the processing in thermal pathways from skin areas generally (Dostrovsky and Hellon, 1978). In numerous neurons of the dorsal horn either static or dynamic responses from the peripheral receptors are extinguished (Hellon and Misra, 1973a). There is a bilateral representation of scrotal receptive fields from the spinal cord until the cortex (Hellon and Mitchell, 1975). At the thalamic level, most of the thermally sensitive neurons in the ventrobasal complex showed a sudden and maintained increase (cf. Fig. 5.10), and others a sudden decrease, for scrotal temperature increases of only 0.5 to 2 °C (Hellon and Misra, 1973b; Jahns, 1977). This behaviour was also found for cortical neurons, but here most of the units suddenly decreased their firing rate with increasing scrotal temperature (Hellon et al., 1973).

5.2 Trigeminal Nucleus

Thermosensitive afferent fibres from the face are connected to second-order neurons in the trigeminal nucleus of the medulla oblongata via the trigeminal ganglion. In the medullary trigeminal nucleus, units responding to cooling the tongue and the face of cats (Poulos et al., 1970; Fruhstorfer and Hensel, 1973; Schmidt, 1976; Dostrovsky and. Hellon, 1978) and monkeys (Poulos, 1971, 1975; Poulos and Molt, 1976) as well as units responding specifically to warming the face of cats (Schmidt, 1976; Dostrovsky and Hellon, 1978) have been found. They are mainly located in the marginal layers of the trigeminal nucleus interpolaris (for details see Darian-Smith, 1973) and the adjacent trigeminal nucleus caudalis. Thermosensitive neurons in this area project to the thalamus (Dostrovsky and Hellon, 1978).

Receptive fields of both cold and warm units are strictly ipsilateral and seem to be concentrated on nose, lips, lower eyelid and pinna (Schmidt, 1976; Dostrovsky and Hellon, 1978). A certain somatotopic

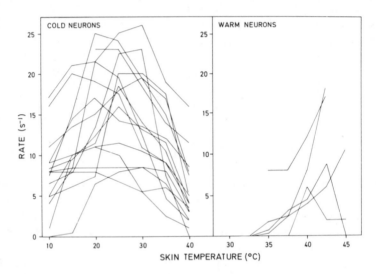

Fig. 5.2. Static firing rate of cold and warm units in the cat trigeminal nucleus as a function of skin temperature. (From Dostrovsky and Hellon, 1978.)

organization in the horizontal plane of the trigeminal nucleus was found. Most receptive fields of cold and warm neurons were between 10 and 100 mm^2 in area, but smaller fields of about 2 to 5 mm^2 and larger ones of 200 to 500 mm^2 were also observed (Dostrovsky and Hellon, 1978). Over half of the thermally sensitive neurons in each group were specifically sensitive to temperature and the remainder had a weak input from mechanical stimulation of the face (Poulos, 1971, 1975; Poulos and Molt, 1976; Dostrovsky and Hellon, 1978).

Cold units in the trigeminal nucleus show a steady discharge depending on the temperature of the face. The discharge is fairly irregular and bears no relation to the burst pattern of peripheral cold receptors, as shown by interval histograms and compound-interval histograms (Schmidt, 1976). However, we should bear in mind that only 25 to 30% of the cat's facial cold receptors show static bursting. As can be seen in Fig. 5.2, the static response curves are similar to those of cold receptors in the nose of the cat. For single neurons, the temperatures for the static maximal frequency may vary between 10 and 35 °C (Fig. 5.3), and some neurons had static maxima at temperatures of 5 °C or below (Schmidt, 1976). If the average static frequencies of cold neurons in the cat's trigeminal nucleus and of cold receptors in the nose are plotted as a function of temperature, and the

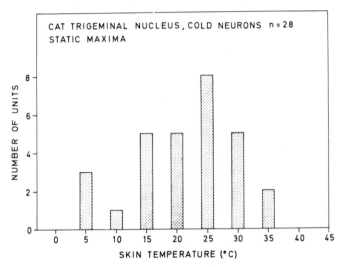

FIG. 5.3. Distribution of static frequency maxima of cold units in the cat trigeminal nucleus as a function of skin temperature. (Data from Schmidt 1976; Dostrovsky and Hellon, 1978.)

higher frequency of trigeminal neurons is accounted for, there is a striking similarity between the response curves of peripheral receptors and central neurons (Fig. 5.4).

From the distribution of the static response of single cold neurons (Fig. 5.3) we may conclude that the across-fibre pattern of peripheral cold receptors (cf. Fig. 4.5) is maintained to a certain extent in the trigeminal nucleus and not extinguished by convergence.

Cold neurons in the trigeminal nucleus respond to cooling steps with an overshoot in frequency and to rapid warming steps with a transient inhibition (Fig. 5.5) in much the same way as found for peripheral cold units in the receptive field. The only difference is that with 5° cooling steps, the ratio between static frequency and dynamic peak frequency of cold neurons (Poulos and Molt, 1976; Schmidt,

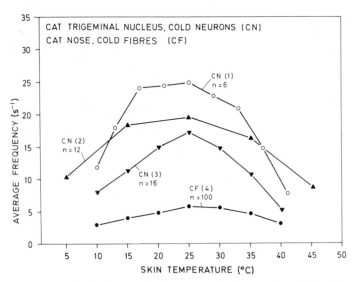

Fig. 5.4. Average static frequency of cold neurons in the cat trigeminal nucleus and of cold fibres in the nose as a function of skin temperature. ((1) From Poulos and Molt, 1976; (2) from Schmidt, 1976; (3) from Dostrovsky and Hellon, 1978; (4) from Hensel and Schäfer, unpublished.)

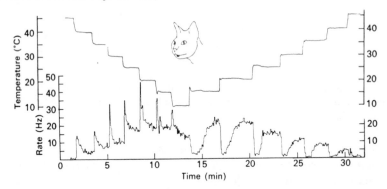

Fig. 5.5 Rate-metre recording (lower trace) of cold unit in the cat trigeminal nucleus whose receptive field centre is indicated on face. Upper trace shows the temperature of the thermode. (From Dostrovsky and Hellon, 1978.)

1976) is in the order of 1:2, while it is about 1:10 for peripheral cold receptors. However, the absolute height of both peak frequencies is in the same order of magnitude, the difference being due to the higher static frequency of cold neurons.

Static and dynamic responses of specific warm neurons in the trigeminal nucleus of cats are shown in Figs 5.2 and 5.6. With increasing skin temperature, the static frequency as well as the dynamic overshoot to warming steps increases in a similar way as found for peripheral warm receptors in the cat's nose (Hensel and Kenshalo, 1969; Hensel and Huopaniemi, 1969). As in peripheral warm receptors, the static frequency reaches a maximum at a certain temperature above which the frequency decreases again.

From these results we can conclude that there is only little processing of peripheral thermal activity in the second-order neurons

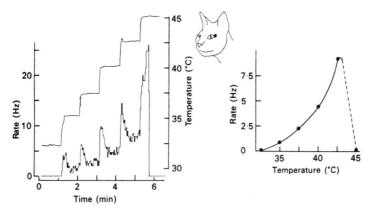

FIG. 5.6. Rate-metre recording of warm unit in the cat trigeminal nucleus whose receptive field at lateral edge of eye is indicated on face. Increasing temperature steps indicated by upper trace in left half of figure. On the right, graph of static firing rate as a function of temperature for the same unit. (From Dostrovsky and Hellon, 1978.)

of the trigeminal nucleus. The specificity of peripheral warm and cold receptors as well as their static and dynamic responses to temperature are preserved. However, there are also some differences: (i) Several peripheral thermoreceptors converge on a central neuron, (ii) the neuronal discharge is irregular and, in case of cold neurons, has a higher frequency than that of peripheral receptors, (iii) burst patterns of cold receptors are not transmitted to the second-order neuron.

As to the neurons in the trigeminal nucleus that respond to both innocuous temperature and mechanical stimulation, their functional significance seems difficult to establish. In the case of cold- and mechanosensitive neurons, it cannot be decided whether they are driven by specific cold receptors and specific mechanoreceptors or whether their input includes afferents from cold-sensitive slowly adapting mechanoreceptors. It has been hypothesized that these bimodal units might provide valuable thermal information (Poulos,

1975), but it remains to be explained why thermally neutral pressure on the skin will not cause a cold sensation, whereas cooling will elicit a sensation of pressure("Weber's deception").

5.3 Thalamus

Units in the ventrobasal complex of the thalamus respond to innocuous cooling of the tongue in cats (Landgren, 1960, 1970; Poulos

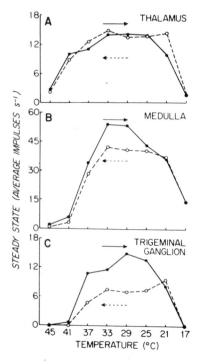

Fig. 5.7. Steady state response as a function of skin temperature for trigeminal cold units at various synaptic levels in the cat. Each point represents the averaged number of discharges during the last 90 s of 3 min test periods following changes in the cooling direction (solid line) and warming direction (dashed line). (From Poulos, 1971.)

and Benjamin, 1968). About 30% of these neurons were specific to cooling and the remainder responded to other qualities of stimulation, such as touch, pressure and taste. Specific thalamic cold units were also described for the trigeminal area in monkeys (Poulos, 1971, 1975; Poulos and Molt, 1976). Reports on specific warm units are rather scarce (Landgren, 1970). Furthermore, in cats there are thalamic

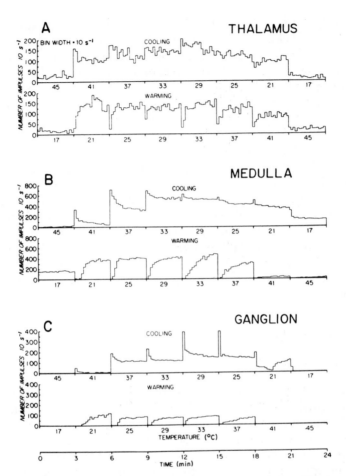

FIG. 5.8. Responses of trigeminal cold units in the cat recorded at different synaptic levels to a sequence of rectangular cold and warm steps, the temperatures of which are indicated on the abscissa. Each stimulus applied to the unit's peripheral field lasted 3 min. The height of each bar indicates the unit's activity expressed as the total number of impulses that occurred in each successive 10 s interval. (From Poulos, 1975.)

neurons excited by innocuous warming and cooling and also by light tactile stimulation of the skin but not by pressure or pinch (Martin and Manning, 1971). The receptive fields ranged from 6 to 96 cm^2, being smaller on the distal extremities and face, larger on the proximal surface of the extremities and face, and largest on the trunk. All units had receptive fields confined to one contiguous area on the contralateral body surface.

All of the units observed had an irregular activity, usually firing in bursts, and the changes in rate during heating were consistent but not always striking. The burst discharge must originate in the thalamus itself because peripheral warm receptors are always firing with a regular sequence of impulses.

Specific thalamic cold neurons show an irregular static discharge at constant skin temperatures and sometimes spindling activity, but it seems unlikely that the burst discharge of peripheral cold receptors is conveyed to this neuronal level (Poulos, 1975). The static discharge frequency depends on skin temperature in a similar way as that of peripheral cold receptors (Fig. 5.7), the only difference being perhaps a larger temperature range of activity in the thalamic neurons, a phenomenon that may be explained by convergence of cold afferents.

Rapid temperature changes of the skin in the trigeminal area elicit dynamic responses of thalamic cold neurones (Fig. 5.8). However, the responses become gradually smaller and disorganized at various synaptic levels from the periphery via the trigeminal nucleus to the thalamus.

5.4 Cerebral Cortex

5.4.1 SINGLE-UNIT ACTIVITY

Even in the somatosensory cortex some specificity of the peripheral thermosensitive input seems to be preserved. Recordings from single cortical cells in the cat's tongue projection area have revealed a number of specific cold units not excited by other stimuli (Landgren, 1957a,b, 1970). These units show an irregular discharge and no pronounced dynamic response to rapid cooling. One specific cortical warm unit was also found. However, there were also cortical cells with a high degree of convergence between various combinations of afferents (Table 5.1). The percentage of neurons responding to only one quality of stimulus seems higher in the ventrobasal complex of the thalamus than in the cortex.

5.4.2 EVOKED POTENTIALS IN HUMANS

When the hand of human subjects is rapidly cooled, evoked cortical potentials can be recorded from the contralateral scalp site that approximates the primary projection of the hand on the postcentral gyrus (Fruhstorfer et al., 1973; Duclaux et al., 1974; Chatt and Kenshalo, 1979). Typical features of the evoked potential were an

TABLE 5.I

Response of thalamic and cortical cells to different types of tongue stimuli in the cat

Effective tongue stimuli	Number of responding	
	Thalamic cells	Cortical cells
Unimodal		
Touch	32	32
Stretch or pressure	4	29
Cooling	20	12
Warming	1	1
Taste	1	5
	58	79
Multimodal		
Touch and cooling	4	8
Stretch and cooling		9
Touch, cooling, and warming		3
Stretch, cooling, warming, and taste		3
Stretch, cooling, and taste		1
Stretch and taste		1
Touch and taste		2
	4	27
Total	62	106

From Landgren (1970).

initial positive peak of large amplitude (25 to 36 μV), long peak latency (325 ms), long duration (200 to 325 ms) and it was graded to the intensity of cooling. On the ipsilateral scalp site, the potential changes were considerably smaller than on the contralateral scalp (Fig. 5.9A). The cold evoked potentials could only be elicited at adapting skin temperatures around 30 °C, whereas at skin temperatures of 40 °C no potential change was seen with a 10° cooling pulse (Chatt and Kenshalo, 1979). From this finding it was concluded that the cortical potential evoked by cooling the skin depends on the activity of specific Aδ cold fibres and not on the discharge of cold-sensitive mechanoreceptors, the latter having their maximal dynamic sensitivity at adapting temperatures of 40 °C.

Rapid warming of the human hand from an adapting temperature of 35 °C also evoked a response at the contralateral parietal scalp site (Chatt and Kenshalo, 1977). This evoked potential had peak latencies from 280 to 356 ms (Fig. 5.9B). Warm stimuli starting at 30 °C adapting temperature were ineffective.

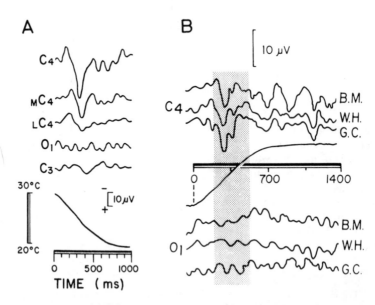

FIG. 5.9. Responses recorded from human scalp evoked by thermal stimuli applied to thenar eminence. A: Response to cooling; $_MC_4$, C_4, $_LC_4$, parietal derivations on right side, each electrode separated by 2 cm from medial to lateral position; C_3, parietal derivation on left side; O, occipital derivation on left side. (From Duclaux et al., 1974.) B: Response to warming (35 to 43 °C). (From Chatt and Kenshalo, 1977.)

It can be assumed that warm evoked potentials have their origin in specific warm primary afferents. Cold evoked cortical potentials may result mainly from the excitation of specific cold receptors served by small Aδ fibres (Chatt and Kenshalo, 1979). Since localized warm, cool and tactile evoked responses following hand stimulation can be recorded at virtually identical scalp sites, it suggests that the topographical organization of the cortex in humans is as valid for the submodalities of temperature as it is for touch and deeper sensibilities (Chatt and Kenshalo, 1977).

5.5 Afferent Connections to the Hypothalamus

At present relatively little is known about the pathways through which spinal and cutaneous thermal afferents are transmitted to the preoptic area and the hypothalamus. Transection experiments (Wünnenberg and Brück, 1970) have indicated that the interruption of the spinal anterolateral tracts abolished or reduced thermoregulatory effector responses to spinal thermal stimulation in guinea pigs. The same conclusion was drawn for the transmission of spinal cold-sensitive afferents in cats (Simon and Iriki, 1971a,b). The relevant spinal temperature signals were apparently conducted in ascending neurons running in these tracts.

On the one hand, afferents from cutaneous thermoreceptors are conveyed via the anterolateral (spinothalamic) tracts and ascend through the lateral lemniscus to the specific nuclei of the ventrobasal complex of the thalamus. On the other hand, we can assume that thermal afferents from skin and spinal cord are relayed to the midbrain raphe nuclei (Weiss and Aghajanian, 1971; Jahns, 1976, 1977) and to the reticular formation via the spino-reticular bundle (Pompeiano, 1973; Albe-Fessard and Bessou, 1973). Various connections between these areas and the hypothalamus have been found by anatomical, autoradiographic and fluorescence microscopic studies (Nauta and Kuypers, 1958; Ungerstedt, 1971; Robertson *et al.*, 1973; Bowsher, 1975), but there is little evidence about the pathways in which thermal afferents are conveyed to the preoptic area and the hypothalamus. One possibility might be a connection from the midbrain raphe nuclei via the central grey matter (substantia grisea centralis) and the ascending fibres from the reticular formation to the hypothalamus, another possibility could be an extralemniscal connection from the mesencephalic reticular formation via the reticulothalamic fibres to the non-specific medial thalamic nuclei (e.g. centrum medianum and nucleus parafascicularis; for references and nomenclature see Albe-Fessard and Bessou, 1973) and from this area to the hypothalamus (Jahns, 1977; Boulant and Demieville, 1977; Brück, 1978a).

Warm-sensitive afferents from the scrotal region of the rat were found to reach the midbrain raphe nuclei, in particular the nucleus medianus raphe, the central grey matter, the unspecific thalamus, and the hypothalamus (Jahns, 1975, 1976, 1977). There was a marked difference in the response curves of these thermoafferent neurons in the rat at various levels of the central nervous system (Fig. 5.10). While the peripheral warm receptors in the scrotal skin showed a typical steady increase in frequency with temperature (Hellon *et al.*, 1975), various populations of neurons were found in the mesencephalic nucleus medianus raphe (Jahns, 1976, 1977), one responding to scrotal warming in a similar way as did the cutaneous warm receptors (type A), whereas the other population (type B) responded with a step-like increase in frequency when the peripheral temperature exceeded 37 °C. In the central grey matter, in the unspecific thalamus, and in the anterior hypothalamus nearly all the units belonged to type A (Jahns, 1977), while in the specific thalamus and the cortex

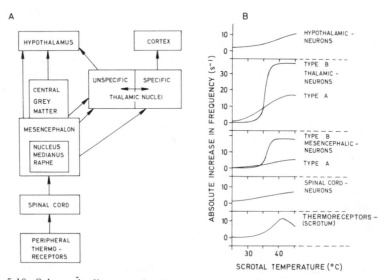

FIG. 5.10. Schematic diagram of pathways for warm afferents from the rat scrotum. A: Suggested pathways; B: average static discharge frequencies of different types of neurons as a function of scrotal temperature. For further explanation see text. (From Jahns, 1977.)

practically all neurons were type B units (Hellon and Misra, 1973b; Hellon *et al.*, 1973).

A further characteristic feature of warm-responsive neurons in the mesencephalic raphe nuclei, the central grey matter and the anterior hypothalamus was the occurrence of periodic oscillations in discharge frequency. At about 30 °C scrotal temperature, the frequency of oscillations was in the order of $1\,s^{-1}$, while at temperatures above 38 °C the frequency was about $4\,s^{-1}$ (Jahns, 1977).

However, it is not known whether these findings hold only for a specialized thermosensitive pathway from the rat's scrotum, or whether they reflect some general properties of neural connections in the thermoregulatory system.

From studies in rats with surgical lesions of hypothalamic connections, Gilbert and Blatteis (1977) concluded that cold afferents may be connected to the preoptic area and the anterior hypothalamus (POAH) from which efferents are mediated via the medial forebrain bundle to the pressor area of the medulla that controls vasoconstriction (cf. Fig. 10.1). A second pathway of cold afferents to the posterior hypothalamus was assumed for the control of shivering. Afferents from warm receptors were thought to be connected with some extra-hypothalamic sites which initiate cutaneous vasodilatation.

Fluorescence histochemistry shows that noradrenaline (NA) and 5-hydroxytryptamine (5-HT) are confined almost entirely to nerve endings in the hypothalamus. Dahlström and Fuxe (1964) found that nearly all the 5-HT-containing cell bodies were localized in the various raphe nuclei with their axons passing caudally to the spinal cord rostrally to the median forebrain bundle to many regions including the hypothalamus. No cell bodies containing 5-HT were seen in the hypothalamus itself. The catecholamine-containing cells were more widely scattered through the brain stem but the most dense concentration was in the mesencephalon at the level of the nucleus interpeduncularis. Many of the fibres from these cells pass up in the median forebrain bundle to the hypothalamus and catecholamine nerve terminals have been found in three hypothalamic areas: preoptic region ventral to the anterior commissure, the supraoptic nuclei, and the paraventricular nuclei (Carlsson et al., 1962; Fuxe, 1965). A few scattered catecholamine cell bodies can also be seen near the third ventricle (Dahlström and Fuxe, 1964; Björklund and Nobin, 1973).

Although this fluorescence microscopy has been conducted primarily in rats, there are also some studies in cats (Pin et al., 1968; Maeda et al., 1973) and human fetuses (Nobin and Björklund, 1973) which show a broadly similar distribution of monoamine neurons. These findings suggest that neurons containing catecholamines and 5-HT if involved at all seem to be concerned with relaying thermal afferents to the hypothalamus, rather than with integration within the hypothalamus or with efferent fibres leaving it.

6
Comparison of Approaches to Cutaneous Thermoreception

6.1 Thermal Afferents and Sensation

6.1.1 FUNCTIONAL SIGNIFICANCE OF CUTANEOUS THERMORECEPTORS

It is difficult to correlate sensory or thermoregulatory events in man with afferent activity from cutaneous thermoreceptors since our knowledge of neural processes is largely based on uncertain analogies from animal experiments. But even if afferent impulses can be directly recorded from human nerves, the data might be biased by fibre selection due to the use of microelectrodes (Järvilehto, 1977).

The existence of neurophysiologically defined warm and cold receptors alone would not be sufficient to prove that they are actually involved in temperature sensation, behavioural responses, and autonomic thermoregulation. In a certain range of warm stimuli, there is no other known kind of cutaneous afferent nerve activity than that from cutaneous warm receptors, and, therefore, their significance can be postulated by exclusion of other possibilities. At lower temperatures, specific cold receptors as well as slowly adapting mechanoreceptors may be excited but there is evidence that the latter are not the neurophysiological basis of cold sensations. Of course, this does not necessarily mean that the cutaneous temperature sensors which are involved in thermoregulatory responses must be the same as those subserving the sensations of heat or cold (Kerslake and Cooper, 1950; Bligh, 1973). On the other hand, there is no evidence for such functional differentiation in the periphery, and it might as well be that the different functions are a matter of central connections and information processing.

According to our present knowledge, warm and cold sensations are mediated by two different systems, one being specific for the quality of

80

warmth and the other for the quality of cold. Blix (1882) has already shown that warm sensations can be elicited only from warm spots and cold sensations only from cold spots. Warming a cold spot was either ineffective or, at higher temperatures, led to paradoxical cold sensations. This specificity has been confirmed with modern methods in human subjects where electrophysiologically defined warm spots gave rise to warm sensations when excited with small warm stimulators (F. Konietzny and H. Hensel, unpublished).

In a series of indirect experiments, the activity of cold fibre populations in monkeys was compared with sensation in human subjects (Darian-Smith et al., 1973; Johnson et al., 1973; Darian-Smith et al., 1975). The results imply that only cold receptors, but neither warm receptors nor slowly adapting mechanoreceptors can account for human cold discrimination. On the other hand, warmth discrimination cannot be explained on the basis of cold fibre inhibition alone but requires an additional population of warm receptors.

Furthermore, reaction times to thermal pulses in human subjects have revealed lower conduction velocities for warm than for cold sensations (Fruhstorfer et al., 1972). Finally, in experiments with nerve blocking, selective inhibition of either cold or warm sensibility has been achieved (Torebjörk and Hallin, 1972; Fruhstorfer et al., 1974).

6.1.2 NEURAL CORRELATE OF STATIC TEMPERATURE SENSATION

While the average discharge frequency of some warm fibre populations depends monotonically on temperature up to the noxious range, the average frequency of large cold fibre populations has a maximum at temperatures above 25 °C and decreases again at lower temperature levels. Because of this bell-shaped curve, the mean static frequency gives ambiguous information of constant temperatures.

Therefore physiologists have searched for temperature-dependent neural parameters other than average frequency. A promising candidate seemed to be the burst discharge whose parameters change in a monotonic way with static temperature between 35 and 20 °C (Iggo and Iggo, 1971; Dykes, 1975; Iggo and Young, 1975; Poulos, 1975; Bade et al., 1979) and thus offer a theoretical possibility of unambiguous information about static temperatures within this range.

However, it is doubtful whether the burst discharge has any significance as an additional neural code. (1) In humans, only few cold fibres, if any, seem to be bursting under static conditions. (2) Neither the average frequency nor the burst parameters of cold fibres from the monkey's hand correlate with static temperature sensation in human subjects. As can be seen in Fig. 6.1, a high static differential sensitivity coincides with a low temperature coefficient of the burst parameters and vice versa. Below 20 °C, the burst discharge disappears but the intensity of cold sensation still increases. (3) There is no evidence that in a system with convergence the burst parameters are conveyed to the second-order neuron, the activity in higher parts

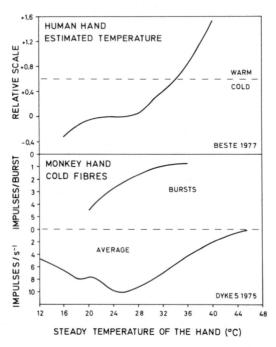

FIG. 6.1 Comparison between estimated temperatures in human subjects and parameters of the static discharge of cold fibres in monkeys as function of steady temperatures applied to the hand.

of the thermal pathway reflecting more or less the static average discharge frequency of peripheral cold receptor populations (cf. Figs 5.4 and 5.7). This argument holds also for any thermoregulatory response elicited from cutaneous cold receptors.

In the temperature range between 34 and 25 °C, the average discharge frequency of cold fibres in the monkey's hand correlates quite well with static cold sensation in humans. The plateau of cold sensation between 27 and 23 °C would correspond to the range (26 to 23 °C) where the average discharge rate of cold fibres remained fairly constant (Dykes, 1975). A possible neural correlate of static sensations in the low temperature range may be an across-fibre pattern including the activity of low-temperature receptors (cf. Fig. 4.5). These receptors may account for static cold sensations below 20 °C (Duclaux et al., 1980).

6.1.3 NEURAL CORRELATE OF DYNAMIC TEMPERATURE SENSATION

For rapid temperature changes there is a fairly good correlation of thermal sensations with neural responses. The difference limen of sensation for paired cooling steps of various magnitude corresponds well with the neural events following the same temperature changes in monkeys, the cumulative impulse count during 3 s after the onset of stimulus being used as an index (Johnson et al., 1973; Darian-Smith et al., 1973, 1975).

Similar comparisons were made between the difference limen of warm sensations and the discriminable stimulus increment (DSI) of warm fibre responses in monkeys (Darian-Smith et al., 1979a,b; Johnson et al., 1979). Figure 6.2 shows an example for paired warming pulses of nearly rectangular shape and various magnitude applied to the palm of the hand in monkeys and humans. The DSI values are calculated for various numbers of warm fibres in a population, using an arbitrary criterion for DSI based on the standard variation of the discharge. The difference limens for trained subjects are estimated by use of a thermode of 1.1 cm^2, the criterion being a 0.75 probability of correct judgments. A certain degree of mismatch between DSI and difference limen as function of stimulus magnitude seems not surprising because neither the decision process nor the central nervous events following warm receptor stimulation are sufficiently known. Moreover, there might also be some species differences between man and monkey.

When applying thermal stimuli of equal magnitude and various rates of change between 0.4 and 5 °C s^{-1}, the magnitude of sensation is practically independent of the rate of change (cf. Fig. 3.3), whereas the peak frequency of the neural discharge clearly increases with the rate of change (cf. Fig. 4.8). This suggests that instantaneous frequency is not a satisfactory correlate of dynamic temperature sensation but that the number of impulses integrated over a certain period of time must also be considered (Hensel, 1952a; Järvilehto, 1973; Kenshalo, 1976). For linear temperature changes, the total number of impulses during the period of change can be linearly related to the magnitude of the stimulus (cf. Fig. 4.8).

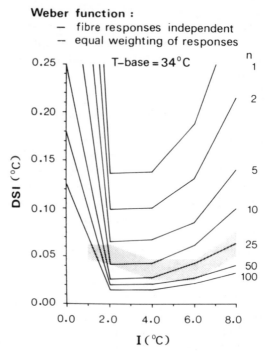

Fig. 6.2. Families of Weber functions relating the discriminable stimulus increment (DSI) for a population of responding warm fibres in monkeys to the mean stimulus intensity (I) of the warming pulses being differentiated. Each curve describes the Weber function for a population of n fibres, where $n = 1, 2, 5, 10, 25, 50$, and 100. T-base was 34 °C. In constructing the responses of each population of fibres equal weighting was given to each fibre's contribution. Each fibre in the population was assumed to respond entirely independently of its neighbours. The response measure was the cumulative impulse count during the first 2 s of stimulation. The shaded zone defines the best observed intensity resolution (difference limen), estimated in trained human subjects for stimulus sequences similar to those used in the fibre studies. (From Johnson et al., 1979.)

However, with equal magnitude of stimulation, slow temperature changes produce a larger total number of impulses than do fast changes. In order to correlate this finding with the facts of sensation, Kenshalo (1976) suggested a "central integrator" with a long time constant of decay. The time constant must be such that the amount of decay is equal to the difference between the total impulses produced by fast and slow temperature changes. Such a system may also account for the fact that the adaptation of thermal sensation following a temperature step is much slower than the adaptation of the receptor discharge. While the adaptation of cold sensation has a half-time of about 4 s (Kenshalo, 1976), the corresponding value for cold fibres is 0.2 to 0.8 s (Hensel, 1952a).

When small areas of stimulation and small temperature changes are used, cold and warm receptors may be excited without conscious thermal sensations. The latter are only observed when a certain number of impulses per unit of time reaches the central nervous system. Attempts have been made to determine this "central threshold" (Hensel, 1952a) for single cold spots in human subjects by comparison with the activity of cold fibres in the cat under identical conditions (Järvilehto, 1973). The threshold of conscious cold sensation would correspond to an instantaneous frequency of 80 s^{-1} or to a total number of 120 impulses in a single cold fibre. Since the cold sensation persists in spite of a decrease in impulse frequency, the "central threshold" seems thus to depend not only on the instantaneous frequency but also on the number of impulses integrated over a certain period of time. The model of a central integrator with a long time constant of decay may give an adequate description of the observations.

When stimulating a single warm spot in human subjects with a small thermode, starting from 35 °C with a slope of 0.8 °C s^{-1}, the average threshold of warm sensations is 37.8 °C. This would correspond to an average instantaneous frequency of 9 s^{-1} or to a total number of 28 impulses in the human warm fibre serving the warm spot (F. Konietzny and H. Hensel, unpublished). The total number of impulses from human single warm receptors at the threshold of conscious warm sensation varies considerably with the rate of change of the warm stimulus (Fig. 6.3), the number of impulses being much higher at slower rates. This corresponds with experiments in monkeys (Kenshalo, 1976).

6.2 Operant Behavioural Estimation of Thermal Sensitivity

Behavioural measurements of the thresholds for thermal stimulation of the face of cats have revealed a temperature sensitivity comparable with that of the human forearm (Kenshalo et al., 1967; Brearley and Kenshalo, 1970). When the thresholds (ΔT) were plotted as a function of adapting temperature (T), and the larger surface area in human experiments was accounted for, the curves for cat and man were surprisingly similar. In contrast to the face, other regions, such as back and thigh, were less sensitive to cooling and highly insensitive to warming, the cats responding only to noxious heat (Kenshalo *et al.*, 1967; Kenshalo, 1968).

Thresholds to warming and cooling stimuli presented to the shaved skin of the inner thigh of rhesus monkeys were measured by the conditioned suppression method at adapting temperatures ranging from 28 to 40 °C (Kenshalo and Hall, 1974). Figure 6.4 shows the behavioural thresholds in the monkey as function of adapting skin

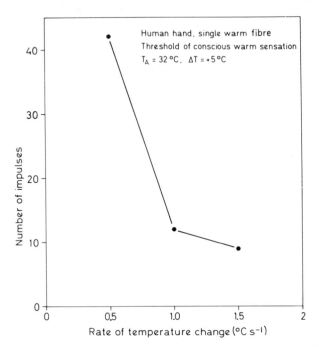

Fig. 6.3. Direct comparison between conscious warm sensation and discharge of a single warm fibre in a human subject. Local warm stimuli of $\Delta T = +5$ °C and various rate of change were applied to the receptive field at adapting temperatures (T_A) of 32 °C. The number of impulses corresponding to the warm threshold is plotted as a function of rate of temperature change. (From Konietzny and Hensel, unpublished.)

temperature. For comparison, human warm and cold thresholds as well as the relative response magnitude of warm and cold fibre populations in rhesus monkeys for a standard thermal stimulus are included in the graph. As can be seen, rhesus monkeys are more sensitive to temperature changes than are human subjects.

Another method for the assessment of thermal sensitivity in monkeys is the measurement of behavioural reaction times (Beitel *et al.*,

1977). When warm stimuli were applied to the upper hairy lip, the monkeys reponded to an increase in the rate of change from 0.3 to 5 °C s^{-1} with a marked decrease in reaction time. In contrast to the effect of the rate of change, the magnitude of the stimulus ($\Delta T = 0.4$ to 5.0 °C) had practically no influence on the reaction time. In addition, the reaction times were shorter when the adapting temperature for the warm stimuli was increased.

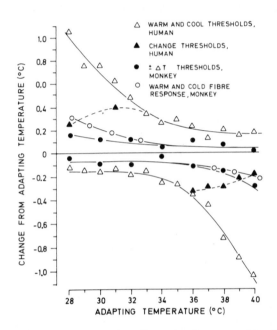

FIG. 6.4. Thresholds for warming and cooling stimuli measured by the conditioned suppression method at seven adapting temperatures between 28 and 40 °C. Warm, cool, and change thresholds measured on the human forearm are also shown for comparison. (From Kenshalo and Hall, 1974; data for warm and cold fibre responses from Kenshalo, 1976.)

It is difficult to interpret these results in terms of sensory quality and neural activity since neither the monkey's sensation nor the type of receptors involved is known with certainty. The cue for the behavioural response may be a change in warm sensation as well as a change in cold sensation, and the cutaneous afferents may originate from specific warm receptors, specific cold receptors, or cold-sensitive mechanoreceptors. As can be concluded from recordings of warm fibre activity in the same experimental situation (Beitel et al., 1977) and from comparisons with other investigations of warm receptors, the behavioural response may depend on a certain impulse frequency, or a certain change in frequency of specific warm fibres. The higher the rate of temperature change, and the higher the adapting temperature, the sooner a certain frequency level will be reached, and the shorter will be the reaction time. This assumption would also explain the fact that the reaction time is independent of the magnitude of the temperature increment. At lower adapting temperatures of 25 °C, the effect of small warm stimuli may depend on an inhibition of cold fibre

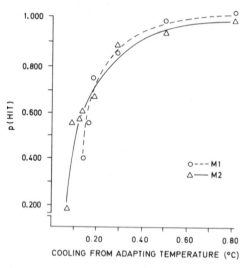

Fig. 6.5. Cool sensitivity in two rhesus monkeys measured by a signal detection method. The proportion of hits, p(HIT), as a function of cooling stimulus intensity (at 1 °C s^{-1}) from a 33 °C adapting temperature at an *a priori* probability of stimulation equal to 0.50 for subjects M 1 (O) and M 2 ($\triangle$). (From Molinari *et al.*, 1976.)

activity rather than on an activation of warm fibres, because in this temperature range the warm fibres are silent.

Using a signal detection method, the cool sensitivity in monkeys was measured by Molinari *et al.* (1976). The monkeys were trained to respond with "yes" or "no" to cold stimuli starting with a rate of change of −1 °C s^{-1} from an adapting temperature of 33 °C. Variables were the magnitude of the stimuli between 0.08 and 0.8 °C and their *a priori* probability, the latter being 0.25, 0.50 and 0.75. Response variables were the proportion of correct positive (hits) and of false positive (false alarms) responses. From these data a detectability index was calculated that increased with the magnitude of the cold stimulus. A classical psychometric function may be derived from the results by plotting the proportion of hits at a stimulus probability of 0.50 as a function of stimulus intensity. This is shown for two monkeys in Fig. 6.5. If the threshold is defined as a 0.50 probability of hits, the value turns out to be 0.19 °C for one monkey and 0.12 °C for the other monkey. This is in fairly good agreement with other behavioural threshold estimations in monkeys (cf. Fig. 6.4). Although the signal detection method is applicable to both human and infrahuman species, whereby the motivational and criterion biases as well as the response biases of the subjects can be accounted for, no direct comparisons between monkeys and human subjects are as yet available.

The role of cutaneous thermoreception for autonomic and behavioural thermoregulatory responses is discussed in Chapters 9, 12 and 13.

7
Central Nervous Thermoreception

Our knowledge of thermoreception in the central nervous system is largely based on animal experiments. Particular problems arise when we try to apply this knowledge to humans, since considerable species differences exist in thermoregulatory functions. For example, sweating responses, as seen in humans, are lacking in many mammalian species. On the other hand, panting and non-shivering thermogenesis are typical for various mammals but do not occur in adult humans.

Other problems arise from the method of thermal stimulation. Local heating of a site in the central nervous system will cause a highly inhomogenous temperature field that is hardly comparable with the rather homogenous field under natural conditions. A more homogenous temperature field over a certain area can be achieved by implanting a grid of several water-circulated metal tubes (Hammel et al., 1963; Jessen, 1976). However, there are still considerable axial temperature gradients along the perfused tubes, and the application of the device is confined to certain areas, such as the hypothalamus of larger animals. As yet, no satisfactory technical solution to the problem of generally applying uniform thermal stimuli to a restricted central nervous area has been found (Simon, 1974).

7.1 Preoptic Area and Hypothalamus

7.1.1 THERMOREGULATORY EFFECTS OF ARTIFICIAL TEMPERATURE VARIATIONS

It has long been known that local thermal stimulation of certain brain areas, in particular the hypothalamic region, will elicit thermoregulatory responses in various species of mammals (for historical references see Bligh, 1973). Since any biological process is dependent on temperature, it is not easy to decide whether a certain neural tissue

90

can be defined as "thermosensitive". In this connection, the localization of the sensitive area, the nature of the response, and the quantitative relationship between thermal stimulus and response has to be considered. As the methods of stimulation became more refined and local thermal probing in unanaesthetized animals (e.g. rats, guinea pigs, cats, dogs, goats, oxen and monkeys) was used as a routine procedure, the concept that thermosensitive neurons are concentrated in the preoptic-anterior hypothalamic (POAH) region developed.

By warming the anterior hypothalamus of unanaesthetized mammals in a thermally neutral environment, it is possible to activate vasodilatation, polypnoea, salivation and sweating. Local heating of the anterior hypothalamic region also reduces cold-induced vasoconstriction and shivering and can activate panting in a cold environment. Hypothalamic cooling in conscious mammals activates vasoconstriction, metabolic increase and shivering at neutral environmental temperatures; it may also inhibit panting and vasodilatation and even activate shivering in a hot environment (for references see Hammel, 1968; Bligh, 1973; Hensel, 1973b).

In unanaesthetized dogs, heating the POAH region increased lingual blood flow in connection with the initiation of panting (Krönert and Pleschka, 1976; Pleschka et al., 1979). The sweat rate in monkeys rose in a linear relation with hypothalamic heating (Smiles et al., 1976). Cooling the POAH region in conscious sheep elicited visible shivering and increased oxygen consumption; blood flow rate decreased in skin and increased in respiratory and non-respiratory muscles, fat and the myocardium; the increase in cardiac output and its redistribution was similar to that which occurs during exposure to a cold environment (Hales et al., 1976, 1977).

Furthermore, non-shivering thermogenesis could be elicited by cooling the anterior hypothalamus of the guinea pig (Brück and Schwennicke, 1971) and the rat (Banet et al., 1978a); on the other hand, warming the POAH region completely suppressed non-shivering thermogenesis induced in the guinea pig by external cooling (Brück and Wünnenberg, 1970) as well as by cooling the spinal cord in the rat (Banet et al., 1978a). Local cooling of the POAH area in goats may also release protein-bound iodine from the thyroid gland; warming the same region blocked the thyroid response in a cold environment (Andersson et al., 1962, 1965a; Andersson, 1970). The catecholamine excretion was found to increase during local cooling of the anterior hypothalamus in goats (Andersson et al., 1963) and pigs (Baldwin et al., 1969). Furthermore, the increase in plasma antidiuretic hormone that occurs in pigs in warm environments was completely suppressed by locally cooling the preoptic region (Forsling et al., 1976).

In order to evaluate hypothalamic thermosensitivity in conscious goats, Jessen and Clough (1973a) implanted a multithermode generating a fairly homogenous temperature field within the POAH region. The method used was an assessment by feedback signals, plotting the static deviation in core temperature as a function of static hypothalamic temperature. Figure 7.1 shows an example. There is no difference

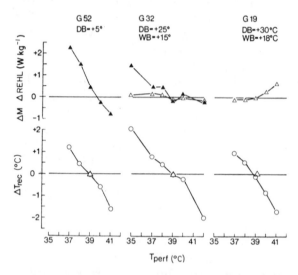

FIG. 7.1. Change in rectal temperature (ΔT_{rec}) of goats after 3 h of hypothalamic thermal stimulation and changes in heat production (ΔM, filled triangles) and respiratory evaporative heat loss (Δ REHL, open triangles) from the control period (1130 to 1200 h) to the first half hour of stimulation period (1200–1230 h) plotted against perfusion temperature (T_{perf}). Five experiments in animal 52 at +5 °C dry bulb (DB), six experiments in animal 32 at +25 °C DB, +15 °C wet bulb (WB), five experiments in animal 19 at +30 °C DB, +18 °C WB. (From Jessen and Clough, 1973.)

in sensitivity and no dead band between responses to cold and warm stimulation, the sensors of core temperature continuously operating even within the narrow range of physiologically occurring core temperatures. Qualitatively, this sensitivity was independent of air temperatures between 5 °C and 30 °C.

Another group of responses that can be elicited by local hypothalamic warming or cooling concerns thermal motivation and behaviour; they are discussed in Chapters 12 and 13.

Recent experiments have shown that the sites of maximal thermo-regulatory responses to warming and cooling the hypothalamus are identical. In the ox, this area was highly localized in the posterior part of the area preóptica (Calvert and Findlay, 1975). In conscious goats with chronically implanted multithermodes, the distribution of thermosensitive sites within the anterior hypothalamus was tested by changes in heat production and respiratory evaporative loss in response to discrete temperature stimuli (Jessen, 1976). The most sensitive areas were situated to either side of the midline in those

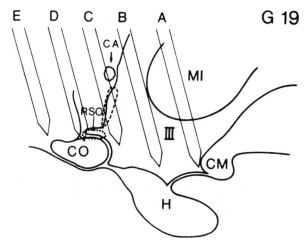

FIG. 7.2. Positions of thermodes in the brain of a goat. Rows A–E projected on midsagittal planes. CA, Commissura anterior; CM, corpus mamillare; CO, chiasma opticum; H, hypophysis; MI, massa intermedia; RSO, recessus supraopticus; dotted lines = nucleus supraopticus, broken lines = nucleus paraventricularis. (From Jessen, 1976.)

frontal planes which contained the nuclei supraoptici and paraventriculares (Fig. 7.2). No difference was found between cold- and warm-sensitive sites.

The question of thermosensitive sites in the posterior hypothalamus was investigated in unanaesthetized squirrel monkeys (Adair, 1974). As Fig. 7.3 shows, there was only very little thermosensitivity in the posterior hypothalamus, as assessed by the method of steady state feedback signals, whereas marked changes in rectal temperature could be elicited by thermal stimuli applied to the proeoptic region. It may be mentioned, however, that marked thermoregulatory behavioural responses could be evoked by thermal stimulation of the

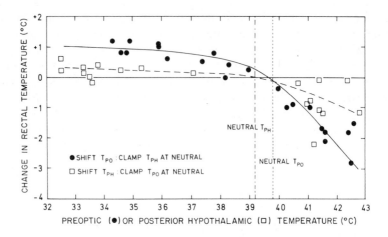

FIG. 7.3. Steady-state changes in rectal temperature of a squirrel monkey at the end of 1-h displacements of preoptic (solid circles, solid line) or posterior hypothalamic (open squares, dashed line) temperature. (From Adair, 1974.)

posterior hypothalamus (p. 197). In conscious goats, cooling the posterior hypothalamus inhibited shivering and increased heat production set up by either external or POAH cooling (Puschmann and Jessen, 1978; cf. p. 126).

7.1.2 ACTIVITY OF THERMOSENSITIVE NEURONS

With microelectrode recordings from the POAH region, various types of units have been identified that respond with positive or negative temperature coefficients to local thermal changes. They are mostly designed "warm-sensitive" and "cold-sensitive" units. In spite of a large amount of data, the results are still difficult to interpret. Is a particular neuron, in fact, functioning as part of the thermoregulatory system and if it is, what role is it playing? Is it a detector of local temperature, an afferent or efferent interneuron? Is the response to warming due to a direct effect of temperature or to an inhibition of cold-sensitive cells and vice versa for the effect of cooling? In contrast to cutaneous thermoreceptors, the central thermosensitive structures have not been identified yet. The dilemma of recording from an unknown structure with an unknown function is reflected by the vague terminology used in the literature, e.g. "thermoresponsive neurons", "thermosensitive structures", "thermodetectors", and the like.

As Eisenman (1972) points out, the electrophysiology of the brain core, including the hypothalamus, is extremely complex as readily isolated pathways or functional centres do not exist. A hypothalamic neuron may be operating in any of the regulatory or behavioural systems known to be influenced by activity in this area of the brain. Further, the multisynaptic nature of the input and output pathways of the brain core makes them sensitive to anaesthetic depression. In general, pentobarbitone, ether, chloralose-urethane, gallamine, and

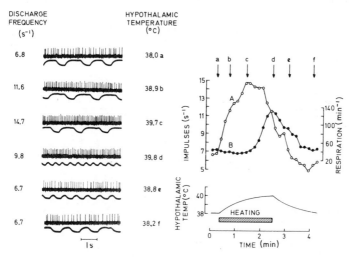

FIG. 7.4. Discharge of a neuron in the preoptic region of a cat (left) and simultaneous record of respiratory rate during local heating of the hypothalamus. Arrows marked with small letters designate time at which the original record was taken. (From Nakayama *et al.*, 1963.)

decamethonium reduced firing rate and usually the temperature sensitivity of the cells, urethane having the smallest depressing effect.

It is widely assumed that changes in neuronal firing induced by thermal stimulations of the POAH region represent neural correlates of the regulatory responses evoked by this stimulus. But even this assumption can be questioned, since thermoresponsive neurons have also been found in the sensorimotor cortex of the cat (Barker and Carpenter, 1970, 1971). Nevertheless, when recording activity in areas influenced by thermal stimulation of the preoptic area or other known thermoresponsive structures, the presumption that thermoregulatory responses are being studied seems reasonable (Eisenman, 1972).

The earliest attempt to detect and identify temperature sensitive neurons in the POAH region of the hypothalamus by recording electrical activities of single nerve cells was made by Nakayama *et al.* (1961). About 20% of neurons in anaesthetized cats had a positive temperature coefficient of their discharge frequency, i.e. responded with an increase in firing rate on local warming (Fig. 7.4). Similar results have been obtained in anaesthetized dogs, guinea pigs and hamsters as well as in unanaesthetized rabbits (for references see Hensel, 1973b; Reaves, 1977; Wünnenberg *et al.*, 1978).

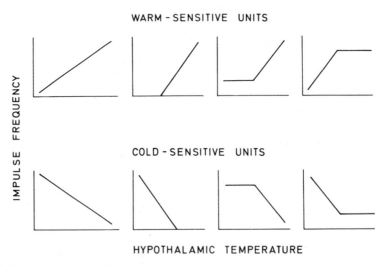

FIG. 7.5. Schematic frequency/temperature curves of the discharge of warm- and cold-sensitive POAH neurons.

Neurons with negative temperature coefficients were found in dogs, cats and rabbits. In the dog, the proportions of insensitive, warm-sensitive, and cold-sensitive neurons were 60, 30, and 10%, respectively, and the three types appeared randomly distributed in the area explored (Eisenman, 1972; Hellon, 1972a). Among thermosensitive hypothalamic neurons in ground squirrels and rats, two basic types were described according to firing rate and ease of recording (Boulant and Bignall, 1973a). A considerably higher percentage of warm-sensitive neurons was found in a group which could be held for longer periods of recording and had higher firing rates.

Some thermosensitive neurons show a linear or continuous relation between firing rate and local temperature over a range of 8 to 10 °C, the most sensitive units often responding with an exponential slope, whereas the others have a non-linear or discontinuous temperature characteristic (Fig. 7.5). These units are thermally insensitive over part of the temperature range, and only above or below a certain threshold level do they show either a warm-sensitive or cold-sensitive response (for references see Guieu and Hardy, 1971; Eisenman, 1972). The sensitivity of the cells with linear or continuous response curves measured on a Q_{10} basis was greater than 2 and reached 15 for the most sensitive units (Eisenman, 1972; Hellon, 1972a; Reaves, 1977). Units with non-linear or discontinuous response curves had a Q_{10} of 2 or less.

By far the most cold-sensitive neurons in the POAH region were of the non-linear type (Guieu and Hardy, 1970b). Since only very few linear, cold-sensitive responses were found (Cabanac et al., 1968), the existence of distinct POAH cold detectors is still undecided. This is not to say that responses to cooling cannot be obtained from this region, since inhibition of static firing of warm detectors could serve this function.

The distribution of thermosensitive neurons in the preoptic area and the posterior hypothalamus of cats (Eisenman, 1972) is shown in Table 7.I). In the ventromedial area of the posterior hypothalamus in rabbits, only 17% of the units responded to local temperature, compared to reports of 30 to 70% for the POAH region. About one-third of the thermally sensitive neurons in the posterior hypothalamus responded only to local thermal stimulation, whereas the remainder could also be excited from the POAH region or the spinal

TABLE 7.I

Distribution of thermoresponsive units in preoptic area and posterior hypothalamus of cats

Neuron type	Preoptic area (%)	Posterior hypothalamus (%)
$Q_{10}1$ units	43.2	47.0
$Q_{10}2$ units	15.9	17.0
High Q_{10} thermodetectors	21.6	7.0
Warm non-linear neurons	7.9	11.0
Cool non-linear neurons	11.4	16.0
Warm-cool units	0.0	2.0

Data from Eisenman (1972).

canal (Wünnenberg and Hardy, 1972). The relative paucity of thermally sensitive neurons in the caudal posterior hypothalamus may be associated with the lower sensitivity of this area to local heating and cooling.

It has been proposed that neurons with linear frequency/temperature characteristics and high thermosensitivity might be primary

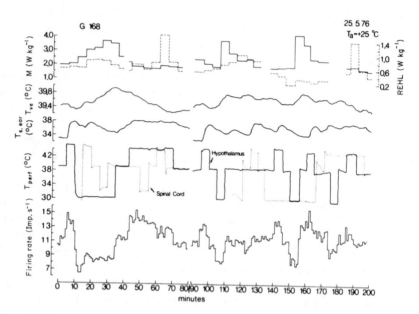

FIG. 7.6. Time course of an experiment demonstrating a warm sensitive hypothalamic neuron in the goat. From top to bottom: Metabolism (M) in Wkg^{-1} (solid line), respiratory evaporative heat loss (REHL) in Wkg^{-1} (broken line), venous blood temperature (T_{ve}), ear skin temperature ($T_{s, ear}$), thermode inlet perfusion temperature (T_{perf}) for hypothalamus (solid line) and spinal cord (broken line), neuronal firing rate (Imp s^{-1}) averaged over 30-s periods. (From Mercer et al., 1978.)

thermodetectors, while units of the non-linear type are likely to be interneurons. This assumption was based on the observation that the linear units were less depressed by barbiturates than the non-linear ones, and that POAH neurons with high thermosensitivity and linear response were not influenced by thermal stimulation of the midbrain and spinal cord, in contrast to non-linear units (Eisenman, 1972). However, as has been shown in rabbits (cf. Fig. 9.9), POAH neurons may change the shape of their response curves with additional inputs from other thermosensitive sites. For example, a unit with linear characteristics can become non-linear, and vice versa (Boulant and Hardy, 1974). Thus it seems problematic to ascribe a certain function to a particular neuron on the basis of its response curve.

Attempts to directly correlate single unit activity in the hypothalamus with thermoregulatory responses have been made in conscious rabbits (Hellon, 1967; Reaves and Heath, 1975; Reaves, 1977) and goats (Mercer *et al.*, 1978). An example is shown in Fig. 7.6. The firing rate of a warm-sensitive neuron in the hypothalamus of a conscious goat follows fairly closely the local temperature change of hypothalamus brought about by a chronically implanted thermode, in correlation with vasomotor and metabolic response, whereas the firing rate of this unit is practically not influenced by simultaneous temperature changes of the spinal cord. Of 53 units, nine were found to be hypothalamically warm-sensitive, two hypothalamically cold-sensitive, and one warm-sensitive to spinal cord temperature changes. While the firing frequency of these 12 neurons only followed changes in local temperature at either hypothalamus or spinal cord, autonomic thermoregulatory responses followed changes in both hypothalamic and spinal cord temperature.

7.1.3 TEMPORAL PATTERNS

Even if a hypothalamic neuron does not change its average frequency, this does not necessarily mean that the unit in question has no thermal sensitivity whatsoever. It was found that the interspike interval distribution may change with temperature, as shown by interspike histograms (Reaves and Heath, 1975). In rats and ground squirrels (*Citellus lateralis*), POAH warm-sensitive and cold-sensitive neurons were found that changed their temperature/frequency curve with time (Boulant and Bignall, 1973b). The most prevalent type was a slow, fluctuating increase and decrease in activity levels but in some units this fluctuation was more rapid, occurring approximately every 10 min. These changes could not be correlated with hypothalamic, colonic, oesophageal, skin, and subcutaneous temperatures. Warm-sensitive neurons in the POAH area in anaesthetized rats showed periodic oscillations of the interspike intervals (Jahns and Werner, 1974; Jahns, 1977). These fluctuations changed significantly with local temperature. Cold-sensitive cells and thermally insensitive cells usually did not reveal a pronounced periodicity. It is not known whether such rhythmical changes can be taken as a relevant parameter of information transmission in the thermoregulatory system.

7.2 Midbrain

Thermosensitive neurons responding to small changes in local temperature have been found in the midbrain reticular formation (Nakayama and Hardy, 1969; Cabanac, 1970; Hori and Nakayama, 1973; Nakayama and Hori, 1973; Hori and Harada, 1976). Upon local cooling, 8% of the units in the midbrain in rabbits at the level of the nucleus ruber and pons responded with high sensitivity (Fig. 7.7)

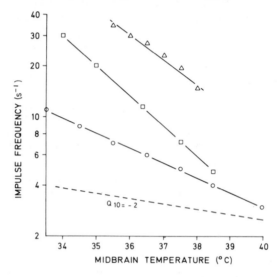

FIG. 7.7. Steady-state responses of three reticular cold-sensitive neurons in rabbits. (From Nakayama and Hardy, 1969.)

and 13% with a non-linear slope (Nakayama and Hardy, 1969). Other units were excited by warming the midbrain (Cabanac and Hardy, 1969). These findings were confirmed by Hori and Harada (1976), who found 18% of the units in the rabbit midbrain reticular formation to be cold-sensitive and 8% to be warm-sensitive.

The physiological significance of these midbrain temperature units is not yet certain. Hardy (1969) has found that cooling the midbrain alone can evoke an increase in oxygen consumption but not so effectively as POAH cooling. Local cooling of the midbrain also induced characteristic bursts of shivering in the unanaesthetized rabbit and a marked increase in the electromyogram showing similar bursts. Local warming of the midbrain immediately inhibited the responses (Cabanac and Hardy, 1969). In recent experiments it was found that altering the local temperature in the midbrain of conscious rabbits had no influence at all on heat dissipation and heat production, while cooling the POAH region induced appropriate thermoregulatory responses (Murakami *et al.*, 1979).

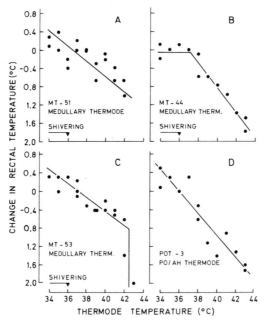

FIG. 7.8. Changes in rectal temperature produced by altering brain temperature for 20 min in rats. A: A proportional relationship between medullary temperature (T_m) and change in rectal temperature (ΔT_r) was noted in about 56% of the animals with effective thermodes. B: No change in T_r with medullary cooling, and a proportional decrease in T_r when T_m was held at neutral and high levels, was seen in approximately 26% of the animals. C: In 18% of the animals, T_r changes that were proportional to T_m except at high-T_m levels were obtained. D: Typical proportional relation between $T_{po/ah}$ and changes in T_r. Lines in A and C fitted by using least-squares method; in B and D, by visual inspection. (From Lipton, 1973.)

7.3 Medulla oblongata

When the temperature of the medulla oblongata of conscious rats was altered, changes occurred in rectal temperature that were generally similar to those produced by altering the temperature of the POAH region (Lipton, 1973). The method of assessing medullary thermosensitivity by feedback signals shows that within certain limits a linear relation exists between rectal temperature and medullary temperature (Fig. 7.8). In some cases the range of proportional medullar thermosensitivity extended from 34 to 43 °C (type A), in other cases there was no change when the medulla was cooled below 37 °C (type B), and in a third group of observations an abrupt fall in rectal temperature occurred with medullary warming above 42 °C (type C). A similar proportionality was found between POAH and rectal temperature in the range between 34 and 43 °C POAH temperature.

TABLE 7.IIA

Response of units in the medulla oblongata of rabbits to changes of medullary temperature

Warm-responsive neurons		39
Linear	22	
Non-linear	17	
Cold-responsive neurons		48
Linear	30	
Non-linear	18	
Insensitive neurons		100
		—
Total		187

TABLE 7.11B

Temperature quotients of warm-responsive neurons in the medulla

Q_{10}	Neurons
2–3	10
3–4	6
4–5	1
5–6	1
6	4
	—
Total	22

From Inoue and Murakami (1976).

As can be seen in Fig. 7.8, the POAH thermosensitivity is only slightly higher than the medullary thermosensitivity. Warming and cooling the medulla oblongata also led to behavioural responses (p. 190).

When lesions were made in the POAH region, the relation between medullary and rectal temperature remained practically unchanged (Lipton, 1973). The medullary thermosensitivity seems thus to be capable of exerting a control of body temperature that is independent

of hypothalamic function. Appropriate thermoregulatory adjustments, such as vasomotor and respiratory responses, were found in
monkeys (Chai and Lin, 1972) and in rabbits (Chai and Lin, 1973)
when the local temperature of the medulla oblongata were altered.

Thermoregulatory and behavioural experiments on the thermosensitivity of the medulla oblongata are in agreement with recordings
from thermosensitive units in the same area in rabbits (Inoue and
Murakami, 1976) and cats (Lee and Chai, 1976). Warm-responsive
and cold-responsive neurons were mainly located in the medullary

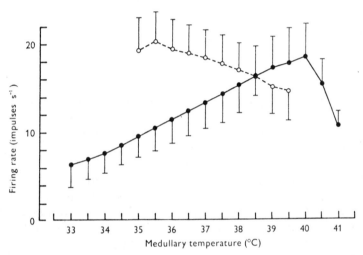

FIG. 7.9. Typical temperature characteristics of five warm-responsive neurons in the
medulla oblongata in rabbits. Mean firing rate was plotted against local temperature
changes. Open circle: Cold-responsive neurons. Filled circle: Warm-responsive
neurons. Bars: Standard error of mean. (From Inoue and Murakami, 1976.)

reticular formation. The majority of these units had a linear tempera-
ture/frequency characteristic over a wide range of medullary temperatures, the warm-sensitive neurons having Q_{10} values between 2 and 6,
in some cases above 6 (Table 7.II). Beyond the range of static
linearity the frequency/temperature curve decreased again, resulting
in a bell-shaped appearance (Fig. 7.9). For the non-linear thermosensitive neurons, the points of intersection of the thermally insensitive part of the characteristics with the warm-sensitive or cold-
sensitive one varied over a wide range of medullary temperatures.

Some medullary warm-sensitive neurons responded to rapid warming with an overshoot in frequency and to cooling with an undershoot,

whereas cold-sensitive neurons responded with an undershoot on warming and an overshoot on cooling (Inoue and Murakami, 1976). It seems an open question whether central thermosensitive neurons can be classified by their static and dynamic properties, the more so as rapid temperature changes are not likely to occur in the central nervous system under natural conditions.

About 50% of the thermosensitive neurons in the medulla oblongata responded also positively and negatively to thermal stimulation of a limited area of the skin (p. 137).

7.4 Spinal cord

7.4.1 THERMOREGULATORY EFFECTS OF ARTIFICIAL TEMPERATURE VARIATIONS

In the last few years it has become evident that all thermoregulatory responses that can be elicited by artificial hypothalamic temperature changes can also be produced by local warming and cooling of the spinal cord in unanaesthetized animals (for references see Simon, 1974). Thermoregulatory responses such as skin vasodilatation and panting as well as vasoconstriction, shivering and piloerection have been observed in dogs (Simon et al., 1963, 1965; Jessen, 1971; Jessen and Mayer, 1971; Jessen and Ludwig, 1971; Jessen and Simon, 1971; Clough and Jessen, 1974; Simon, 1974), pigs (Ingram and Legge, 1971, 1972a,b; Carlisle and Ingram, 1973a), sheep (Clough et al., 1973; Hales and Iriki, 1975; Hales et al., 1975; Bacon and Bligh, 1976), goats (Jessen and Clough, 1973b), oxen (McLean et al., 1970; Jessen et al., 1972), monkeys (Chai and Lin, 1972), rabbits (Iriki, 1968; Kosaka and Simon, 1968a,b; Guieu and Hardy, 1970a; Riedel et al., 1973), guinea pigs (Brück and Wünnenberg, 1967, 1970; Brück et al., 1971), and rats (Lin and Chai, 1972; Banet et al., 1978a). In addition, sweating responses on spinal heating were observed in oxen (McLean et al., 1970) and monkeys (Chai and Lin, 1972). Spinal cooling induced non-shivering thermogenesis in rats (Banet et al., 1978a) while similar responses were not observed in guinea pigs (Brück and Wünnenberg, 1970).

Figure 7.10 shows the thermoregulatory responses of an anaesthetized dog during cooling of the vertebral canal. Although no temperature changes are recorded in brain, rectum and aorta, the animal responds to spinal cooling with shivering, metabolic increase, and vasoconstriction in the hindpaw, as established by a fall in skin temperature. Similar results were obtained in conscious dogs, showing that shivering could be regularly induced by selective cooling within the spinal canal at thermoneutral ambient conditions (Simon *et al.*, 1965).

Two questions arise in connection with these findings. The first question concerns the location of the spinal thermosensitive structures. In dogs with chronical bilateral transection of the dorsal roots

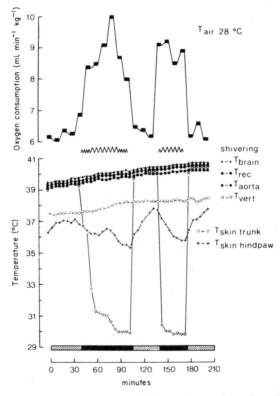

FIG. 7.10. Selective cooling (black bars) within the lower thoracic and lumbar spinal canal of an anaesthetized dog by means of a water-perfused peridural thermode at warm ambient conditons. Shivering and increased oxygen consumption are induced by lowering spinal canal temperature in spite of elevated, rising extraspinal core temperatures. The fall of skin temperature at the paw during spinal cooling indicates cutaneous vasoconstriction. (From Simon *et al.*, 1963.)

of the lumbosacral spinal cord, selective cooling of this deafferentiated section was still effective in evoking cold tremor (Meurer *et al.*, 1967; further references see Simon, 1974). Thus it is most likely that the thermosensitive structures are located in the spinal cord itself. The second question is whether the effects of spinal cooling may be due to unspecific effects on signal transmission in the spinal cord, or whether they can be ascribed to the stimulation of specific thermosensors. A survey of the experimental evidence shows that any kind of thermo-regulatory responses can be elicited from the spinal cord, and that these responses do not differ from those elicited by hypothalamic thermal stimulation or by natural heat and cold stimuli (for references see Simon, 1974).

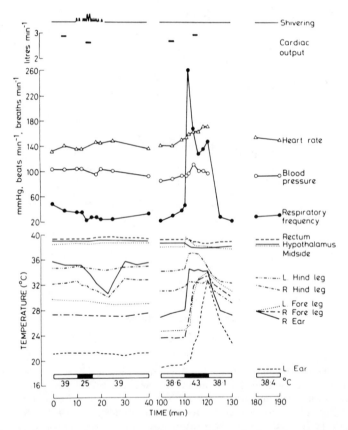

Fig. 7.11. Effects of local temperature changes of the spinal cord on body tempera-tures, cardiovascular, respiratory and shivering activity in a sheep. (From Hales and Iriki, 1975.)

Spinal cord heating in unanaesthetized dogs and sheep elicits panting (polypnoea) and cutaneous vasodilatation (Jessen, 1967; Hales and Iriki, 1975). Figure 7.11 shows various cardiovascular and respiratory responses of a conscious sheep during spinal cooling and warming. These responses occur although rectal and hypothalamic temperatures change in opposite directions as compared with spinal temperature. Spinal cooling leads to shivering and cutaneous vasoconstriction and a decrease in cardiac output whereas spinal heating is followed by an increase in respiratory frequency, cutaneous vasodilatation and increase in cardiac output.

Prolonged spinal cord heating caused a static fall in rectal temperature. However, spinal cord cooling led to an increase in core temperature only at ambients of 5 °C, whereas at 30 °C the internal

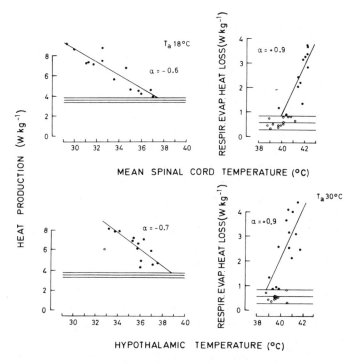

FIG. 7.12. Metabolic heat production and respiratory evaporative heat loss as function of spinal and hypothalamic temperature of a conscious dog at ambient air temperatures (T_a) of 18 and 30 °C, respectively. (From Jessen, 1971.)

cold stimulus was ineffective. It can be concluded that only spinal heating is conveyed through an independent feedback loop, while the transmission of cold signals from the spinal cord is dependent on the presence of cold afferents from the skin (Jessen and Clough, 1973b).

Studying conscious dogs at constant ambient temperatures, Jessen and Mayer (1971) and Jessen and Simon (1971) found that spinal cord and hypothalamus represent equivalent core sensors of temperature. Cooling the spinal cord increased heat production in a reversed linear relation to cord temperature by -0.8 W kg^{-1} °C^{-1}. The respective value for the hypothalamus was the same. Spinal cord heating produced a linear increase in evaporative heat loss by $+1.0$ to $+1.4$ W kg^{-1} °C^{-1}, while heating the hypothalamus was followed by an increase of $+1.0$ W kg^{-1} °C^{-1} (Fig. 7.12). A similar equivalence of spinal cord and hypothalamus has been found for thermal vasomotor responses in the anaesthetized dog (Simon, 1971).

7.4.2 ACTIVITY OF THERMOSENSITIVE NEURONS

Microelectrode recordings from single units in the rhombencephalic part of the medial lemniscal pathway were carried out by Wünnenberg and Brück (1968a,b, 1970) in guinea pigs, in which the lower cervical and upper thoracic segments of the spinal cord were electively heated. These investigations presented the first evidence for transmission of warm-sensitive afferents originating in the spinal cord. The neurons showed static and dynamic sensitivity to warming, and their static frequency increased with spinal temperature in a non-linear way of varying degrees. The average frequency/temperature relation of eight single units could satisfactorily be described by a Q_{10} of about 11 in the temperature range between 38 and 45 °C.

In anaesthetized cats, single unit activity in the peripheral parts of the ascending spinal anterolateral tract was recorded (Simon and Iriki, 1970, 1971a,b). The recording site was at the level of C2 to C5, rostral of the stimulated sections of the spinal cord. In order to prevent influences from descending neurons, which could have been activated by supraspinal structures in response to the spinal thermal stimulation, high spinal transections were performed. When applying static temperatures between 14 and 47 °C, the temperature in the spinal canal varied between 22 and 43 °C. This temperature was considered to be representative as stimulus temperature for the spinal neurons.

Two clearly separated sets of thermosensitive neurons could be distinguished according to their static response to temperature displacements within the spinal cord (Fig. 7.13). A population of warm-sensitive units showed an increase in discharge frequency with an average maximum around 43 °C, while a population of cold-sensitive neurons increased their frequency with falling spinal temperature until a maximum was reached at temperatures between 30 and 25 °C. Below this temperature the discharge frequency decreased again. At static temperatures above 41 °C, a second increase in frequency of the cold-sensitive neurons occurred, comparable to the "paradoxical" discharge of cutaneous cold receptors. The steepest and approximately linear parts of both response curves covered the range between 35 and 41 °C and intersected approximately at 37 °C. Average static sensitivities in this range amounted to 5.6 s^{-1} °C^{-1} for warm-sensitive and -2.4 s^{-1} °C^{-1} for cold-sensitive units. The sensitivities are of the same order of magnitude as those described for hypothalamic heat- and cold-sensitive neurons.

About half of both warm- and cold-sensitive neurons exhibited dynamic components in their responses to temperature changes.

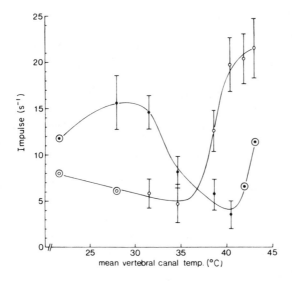

Fig. 7.13. Average discharge frequency of cold- and warm-sensitive neurons in the spinal cord of cats as function of spinal temperature. (From Simon and Iriki, 1971b.)

Furthermore, a high degree of convergence between spinal and cutaneous thermal afferents was observed in the ascending spinal cold-sensitive neurons of anaesthetized spinalized cats (Simon, 1972) but this convergence appeared less obvious in animals with intact connections between the spinal cord and the brain (Simon, 1974). Therefore, definite conclusions as to the degree of convergence of spinal and cutaneous thermal afferents are not yet possible.

8
Extra-central Deep Body Thermoreception

The existence of extrahypothalamic deep body thermosensors has long been maintained. This view was derived from thermoregulatory responses that could not be explained on the basis of hypothalamic and cutaneous temperature alone (Thauer, 1939). Part of these phenomena may now be accounted for by thermosensitivity in lower parts of the central nervous system, such as midbrain, medulla oblongata and spinal cord. But there seem also to be thermosensors outside the central nervous system situated elsewhere in the deeper tissues of the body. At present we have relatively little evidence as to the site and function of such structures.

8.1 Blood Vessels

Some evidence of extracentral temperature sensitivity was reported by Blatteis (1960). He found that when the limb of an anaesthetized dog was isolated from the trunk except by its main blood vessels, cooling the limb resulted in shivering elsewhere in the body, although there was apparently no relation between the onset of shivering and deep body temperature, mean skin temperature or brain temperature. It was proposed that the shivering may have resulted from the excitation of as yet undescribed temperature sensors in or near the large blood vessels. Bligh (1961) found that when cold saline was infused into the abdominal vena cava of the conscious sheep during heat exposure, at a rate which prevented any rise in the temperature of the blood supply to the brain, respiratory frequency was depressed. He therefore suggested that temperature sensitive structures were present in or near the vena cava. This assumption was tested by Cranston *et al.* (1978) in conscious rabbits. Infusions of hot and cold solutions were made into the hepatic portal vein, the inferior vena cava and, for control, into an ear vein. It turned out that the increases in brain temperature were identical when the solutions were given into the ear vein, the portal vein or the inferior cava. Likewise, the decreases in brain temperature did not differ when cold solutions were infused into the ear vein or the vena cava. Only with cold infusions into the portal vein, the responses in brain temperature were slightly but significantly smaller. Thus, there is no evidence for the presence of warm

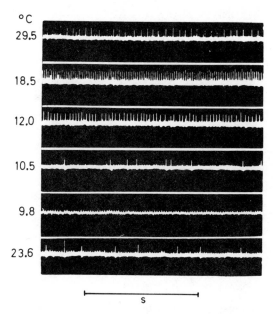

Fig. 8.1. Afferent impulses from a single fibre serving a cutaneous vein in the rabbit during perfusion with solutions of different steady temperatures. (From Minut-Sorokhtina, 1978.)

receptors in the inferior vena cava, portal vein, liver or hepatic vein, and no evidence for a concentration of cold receptors in the inferior vena cava, but there may be cold sensitive elements in the portal vein or the tissue perfused by blood passing through it.

From studies in humans during exercise, the hypothesis was derived that thermoreceptors might be situated in the deep leg veins or possibly in the muscle itself. During muscular work, the local temperatures may increase to 40 °C or higher, and it was thought, therefore, that some local thermoregulatory reflex may be elicited from these regions (Saltin et al., 1968; Gisolfi and Robinson, 1970;

Saltin and Gagge, 1971). However, no direct proof of the existence of muscle thermoreceptors has been obtained as yet. For example, sweat rate is not influenced by high muscle temperatures during positive or negative work (Nielsen, 1969; Nielsen et al., 1972).

Electrophysiological evidence for the existence of thermosensitive structures in the veins has been obtained by Minut-Sorokhtina (1968, 1972, 1977, 1978). She recorded from afferent nerve fibres serving isolated or intact veins in the rabbit's ear when the local temperature

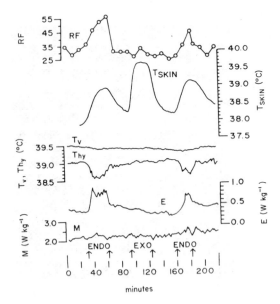

Fig. 8.2. Respiratory evaporative heat loss response to an intra-abdominal heat input of 0.5 W kg^{-1} in the sheep. RF, Respiratory frequency; T_{skin}, temperature of skin surface directly over implanted heat sources; T_{hy}, hypothalamic temperature; E, respiratory evaporative heat loss; M, metabolic heat production; ENDO denotes period of internal heating and EXO indicates external heat application to skin. (From Rawson and Quick, 1970.)

was varied (Fig. 8.1). There was a static discharge, the frequency of which rose when the constant temperature was set at lower levels, and an overshoot on rapid cooling. It was suggested that excitation of the receptive terminals results from contraction of vascular smooth muscle fibres during cooling. Thompson and Barnes (1969) passed a catheter through a segment of the femoral vein in the cat and perfused it with fluid within the physiological range (39 to 41 °C). Action potentials were recorded in the nerve filaments of the 5th dorsal root, showing a rise in discharge frequency when the temperature increased.

It remains an open question whether these thermosensitive structures act as a source of temperature information in the thermoregulatory system. In order to answer this question, it has to be demonstrated that feedback signals originate from the sites in question.

8.2 Abdominal Cavity

Evidence for deep body thermosensitivity in the ewe was given by Rawson and Quick (1970, 1971a, 1976). At neutral environmental

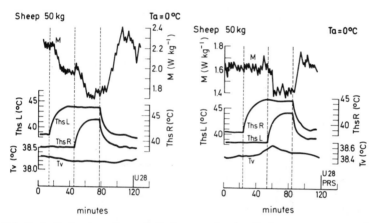

FIG. 8.3. Left: Reduction of metabolic heat production in the ewe upon warming the abdominal viscera by means of implanted electric heat sources. Plotted from the top are the metabolic heat production (M) in W kg^{-1}, the surface temperature of two heaters on the left side (Ths L), two on the right side (Ths R), and vaginal temperature (Tv). Ta, Environmental temperature. Right: Abolition of one half of the total response following right unilateral splanchnotomy in the same ewe. (From Rawson and Quick, 1971b.)

conditions, panting could be elicited by local intra-abdominal heating, although the temperature of the hypothalamus decreased and that of the spinal cord remained constant. Neither can the effect be explained by the simultaneous increase in skin temperature of the abdomen, since separate warming of the skin was ineffective (Fig. 8.2). When intra-abdominal heating was performed at low ambient temperatures, it reduced metabolic heat production, while in hot environments it enhanced panting. Unilateral section of the splanchnic nerves (Fig. 8.3) abolished the response to warming of the ipsilateral side without changing the response on the opposite side (Rawson and Quick, 1971b). Ruminal cooling with water of 16 °C caused an immediate decrease in evaporative heat loss and respiratory frequency, while the metabolic heat production did not increase until the core temperature had declined by about 0.75 °C (Rawson and Quick, 1972).

It was concluded that the thermoreceptors stimulated lie within the walls of the rumen and intestine, and possibly in the mesenteric veins,

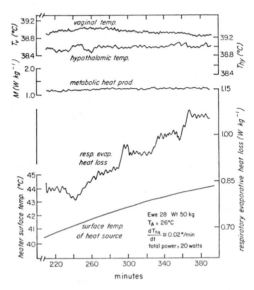

FIG. 8.4. Increase in respiratory evaporative heat loss in steps during slow warming of intra-abdominal thermoreceptors in the ewe. (From Adair and Rawson, 1973.)

the splanchnic nerves being the afferent pathway for these receptors. It seems possible that abdominal thermosensitivity is better developed in herbivorous species and ruminants, where the temperature of the rumen can increase to more than 40 °C due to the active fermentation of cellulose (Trautmann and Hill, 1949). This would be consistent with the absence of evidence for abdominal cold sensitivity in the rabbit (Riedel et al., 1973) and the absence of any response to abdominal heating and cooling in pigs (Ingram and Legge, 1972a).

When the intra-abdominal warm receptors were excited with a slow increase in temperature of 1.2 to 2 °C h^{-1}, the increase in respiratory evaporative heat dissipation at high ambient temperatures (Fig. 8.4) as well as the decrease in metabolic heat production at low ambient temperatures did not occur continuously but in steps of about 40 min duration (Adair and Rawson, 1973). This phenomenon cannot be explained as yet, but it shows that the input/output relations of the thermoregulatory system, at least in this case, cannot adequately be described by a simple continuous function.

In the rabbit a 0.9 °C increase in abdominal temperature was followed by cutaneous vasodilatation (Slepchuk and Ivanov, 1972). An increase in rectal temperature of 0.2 °C was equal to a 1 °C increase in hypothalamic temperature in its effect on vasodilatation, while for panting an increase in rectal temperature of 0.8 °C was equivalent to a 1 °C increase in hypothalamic temperature (Kluger et al., 1973). The thermosensitivity of the abdomen in rabbits was found to be about one-fourth of that of the spinal cord (Riedel et al., 1973). In monkeys, the sensitivity of abdominal thermosensors, as assessed by thermoregulatory responses, was about one-third of hypothalamic thermosensitivity (Kluger et al., 1973). The existence of thermosensitive structures within the abdomen was also established by behavioural experiments in squirrel monkeys (p. 192).

There is also some electrophysiological evidence for intra-abdominal thermosensitive structures. Warm stimulation of the dorsal abdominal wall in rabbits evoked static and dynamic warm-sensitive responses in single-fibre preparations of the splanchnic nerve (Riedel, 1976). The receptive field was located in a circumscribed area on the ipsilateral dorsal abdominal wall extending between the suprarenal gland and the origin of the superior mesenteric artery. The fibres were non-myelinated, with conduction velocities of 0.6 to 1.1 ms^{-1}, and were not excited by mechanical stimulation. At normal core temperature (38 to 39 °C) these fibres discharged with a low rate. Warming the receptive field elicited a dynamic increase in frequency with subsequent adjustment to a steady level, sometimes with periodic oscillations in frequency (Fig. 8.5). Cooling the abdominal wall was followed by a transient inhibition of impulse frequency. Fibres with lower discharge rate (type A) usually had static and dynamic maxima at about 40 °C, while fibres with higher discharge rate (type B) had dynamic maxima near 46 °C. The temperatures of the static maxima agree quite well with those of cutaneous warm fibres, but the discharge rates of splanchnic warm receptors are considerably lower than the frequencies of warm receptors in the external skin.

When the stomach or the adjacent part of the duodenum in cats were cooled by means of a water-circulated metal thermode, afferent impulses could be recorded from single-fibre preparations of each greater splanchnic nerve (Gupta *et al.*, 1979). These fibres were silent at higher temperatures but produced a continuous discharge when a thermode within the stomach was cooled to 12 °C (Fig. 8.6). The assessment of static threshold temperatures was difficult since the data did not allow a precise location of the receptive field. There was no indication of a pronounced dynamic sensitivity, and no response to mechanical stimulation was seen. The receptors that were connected

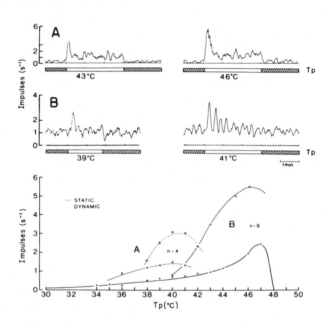

Fig. 8.5. Upper part: Impulse frequency of single abdominal warm fibres in the rabbit (type A and B) during thermal stimulation with different perfusion temperatures (T_p). Hatched bar indicates core temperature (39.1 °C in A, 38 °C in B). Lower part: Average static and dynamic impulse frequency of the two populations of warm fibres. (From Riedel, 1976.)

with non-myelinated fibres were presumably in the stomach or the upper segment of the duodenum.

The results suggest that the effects of ruminal cooling in the ewe (Rawson and Quick, 1972a) and of intragastric cooling in human subjects (Nadel *et al.*, 1970) may not only be due to central cooling of the body, but that local afferents from the stomach may also play a certain role.

Among the mechanosensitive afferents from stomach, intestine, mesentery, and in or around aorta and vena cava, none was sensitive to cooling (Gupta et al., 1979).

Spontaneous burst discharges at temperatures between 32 and 42 °C have been observed during recording from single units in Auerbach's plexus in the cat (Wood, 1970). The interspike interval decreased within this range and the number of spikes per burst increased with rising temperature, the Q_{10} values for the first parameter being 12 to 16 and for the second parameter about 3. These

FIG. 8.6. Cold-sensitive afferent from the greater splanchnic nerve in the cat. Response to intragastric cooling to 12 °C. The receptive field was either in the stomach or upper intestinal part. (From Gupta et al., 1979.)

neurons were thought to be involved in the control of intestinal motility.

8.3 Other Sites

Recordings of afferent impulses from few-fibre and single-fibre preparations supplying the knee joint in cats and dogs have shown that considerable changes in discharge rate occur during cold or warm stimulation (Yamada et al., 1976). Part of the responses to cooling were ascribed to the stimulation of cold-sensitive mechanoreceptors, but there were also afferents that apparently originated in specific cold and warm receptors.

Mercer and Jessen (1978a,b) have assessed the "residual inner body" thermosensitivity, that is, deep body thermosensitivity outside the hypothalamus and spinal cord (p. 124). It is not known from which areas the thermal afferents originate. Besides central nervous sites with established thermosensitivity, such as midbrain and medulla oblongata, extra-central areas may as well be involved.

Lung stretch receptors innervated by vagal afferent fibres usually did not respond to cooling (Golenhofen, 1970; Gupta et al., 1979). However, in the cat one vagal fibre was found which responded to respiratory movements but at the same time had a high sensitivity to lowering the body temperature from 38 to 36 °C. It is not known whether this type of fibre plays any role for thermoreception (Golenhofen, 1970).

Afferents from atrial type A and B receptors were not sensitive to cooling (Gupta *et al.*, 1979). However, as sympathetic activity from the thorax was not yet studied during cooling, the existence of cold-sensitive receptors served by sympathetic fibres cannot definitely be excluded.

Recently it was shown that carotid baroceptors and chemoreceptors of cats are highly sensitive to static temperatures in the physiological range, the discharge frequency of both types having a positive temperature coefficient (Gallego *et al.*, 1979). The static thermal sensitivity $(\Delta F/\Delta T)$ of carotid chemoreceptors, as expressed by the change in discharge frequency (ΔF) for a given change in static temperature (ΔT) is in the same order of magnitude as the static sensitivity of the cat's cutaneous warm receptors and higher than that of hypothalamic thermosensors.

It can be concluded that during hyperthermia the discharge frequency of carotid baroceptors increases considerably and, therefore, arterial blood pressure decreases. This would mean that heat syncope is not only a matter of regulatory centres and effector mechanisms but, at least in part, also a matter of receptor processes. On the other hand, baroceptor reflexes in man can also be influenced by afferents from peripheral thermoreceptors. Reactions of finger blood flow to lower body negative pressure were enhanced when the contralateral hand was cooled, and suppressed when the hand was heated. This effect was assumed to depend on central interaction of baroceptive and thermoreceptive drives (Heistad *et al.*, 1973).

Activation of carotid baro- and chemoreceptors induces not only the classical changes in ventilation and blood pressure, but also affects the hypothalamus (Yamashita, 1977). It is thus likely that increases in blood temperature may also increase production of antidiuretic hormone and reduce urinary flow via a chemoreceptor reflex. We can even further speculate that changes in carotid body temperature may also effect the temperature regulation centres of the hypothalamus and thus act as a feedback loop involved in temperature regulation.

9
Interaction of Thermal Drives

Thermoregulatory responses as well as thermal comfort and behaviour can be induced by temperature stimuli from various sites of
the body. It is well established that a combination of thermal inputs
leads to some kind of integrated response. Various authors have
described by mathematical expressions the signal interaction from
two or more simultaneously stimulated thermoreceptive sites for a
given thermoregulatory response. In fact, a quantitative assessment of
signal interaction has its limitations, since the temperature fields at
the stimulated areas cannot satisfactorily be defined. For a given
control action, the combination of two thermal drives T_1 and T_2 was
found in most cases to be either additive or multiplicative. A clear
differentiation is only possible when a large range of temperatures T_1
and T_2 is investigated; otherwise it is often impossible to assess
whether an interaction is multiplicative or additive. Therefore, the
distinction between the two types of combinations has probably little
importance (Cabanac, 1975).

9.1 Combined Thermal Drives in Animals

9.1.1 ANTERIOR HYPOTHALAMUS AND SKIN

Thermoregulatory effects of warming and cooling the anterior
hypothalamus can be modified by skin temperature (for references see
Bligh, 1973; Hensel, 1973b; Hensel et al., 1973; Cabanac, 1975). In
experiments that first demonstrated thermoregulatory responses to
hypothalamic cooling in conscious animals (Kundt et al., 1957a;
Krüger et al., 1959), vasoconstriction during hypothalamic cooling
could be inhibited by an increase in skin temperature, thus suggesting
an interaction between thermal inputs from the anterior hypothalamus and from cutaneous thermoreceptors.

In conscious dogs, heat production by shivering and heat loss by panting are approximately linearly related to hypothalamic temperature (Hammel, 1970, 1972; Jessen, 1971; Jessen and Ludwig, 1971; Jessen and Mayer, 1971; Jessen and Simon, 1971), the response characteristics depending on skin temperature. Interactions between skin and hypothalamic temperature were also found in conscious pigs with respect to panting and oxygen consumption (Baldwin and Ingram, 1968a; Ingram and Legge, 1972a,b), peripheral blood flow (Ingram and Legge, 1971) excretion of catecholamines (Baldwin et al., 1969) and antidiuretic hormone (Forsling et al., 1976). In conscious sheep either hypothalamic, spinal, or cutaneous warming elicited a marked increase in the proportion of cardiac output passing through cutaneous arteriovenous anastomoses (Hales et al., 1975).

All these interactions between hypothalamic and skin temperature consist in an enhancement of thermoregulatory responses if both temperatures are changed in the same direction, and in a mutual inhibition if the changes occur in opposite directions. According to Hammel (1968), thermoregulatory responses (R) in conscious dogs can be described by the equation

$$R - R_0 = k(T_h + \bar{T}_s - a); \ R \geq 0 \tag{14}$$

where k is a proportionality factor, T_h hypothalamus temperature, $\bar{T}_s$ average skin temperature, and a is a constant. From Eqn (14) an addition of central and peripheral thermoregulatory drives can be concluded. However, in other cases of the same series of experiments, the relationship was of a multiplicative type.

The sweating rate of conscious rhesus monkeys increased linearly with hypothalamic temperature, the slope being $7.44 \ \mathrm{mg \ cm^{-2} \ h^{-1}}$ $^\circ\mathrm{C}^{-1}$ (Smiles et al., 1976). At lower skin temperature, a parallel shift of this characteristic towards higher hypothalamic temperatures but no change in slope occurred. From the available data it cannot be decided whether the interaction of thermoregulatory drives is of an additive or multiplicative type.

In other experiments there was a multiplicative relation between skin and hypothalamic temperature, as expressed by a hyperbolic function. The general equation for this relation takes the form

$$R - R_0 = k(T_h - T_{h,0})(\bar{T}_s - \bar{T}_{s,0}) \tag{15}$$

The symbols are the same as in Eqn (14). There are reference temperatures for both hypothalamic and average skin temperature that are fixed and the same for all thermoregulatory responses. Similar equations were suggested for the thermoregulatory heat production in conscious dogs (Cabanac, 1970) and cats (Jacobson and Squires, 1970) as well as for the rate of respiratory heat loss or heat tachypnoea in dogs (Chatonnet et al., 1964). At hypothalamic temperatures below 40.3 °C panting can be prevented by reduction of skin temperature. At temperatures above 40.3 °C even a great reduction in skin temperature is clearly ineffective in inhibiting panting. Furthermore, at high skin temperatures panting can no longer be inhibited by decreasing hypothalamic temperature.

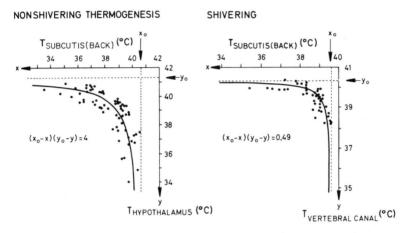

FIG. 9.1. Left: Combinations of subcutaneous and hypothalamic temperature at the threshold of non-shivering thermogenesis in conscious guinea pigs. (From Brück and Wünnenberg, 1967.) Right: Combinations of subcutaneous and spinal canal temperature at the threshold of shivering in conscious guinea pigs. (From Brück and Schwennicke, 1971.)

For non-shivering thermogenesis in guinea pigs as function of hypothalamic and subcutaneous temperature (Brück and Wünnenberg, 1967) and for shivering as function of spinal and subcutaneous temperature (Brück and Schwennicke, 1971), a multiplicative relationship was found, as depicted in Fig. 9.1. The nearly symmetrical shape of the hyperbolas indicates that the thermal drives from skin and body core have approximately equal strength. This seems to be typical for small animals, whereas in larger ones the thermal drive from the body core is more intense than that from the skin.

In conscious rabbits respiratory frequency and metabolic heat production were found to depend approximately linearly on the local temperature of the anterior hypothalamus. Changes in skin temperature altered the threshold as well as the slope of the response curves,

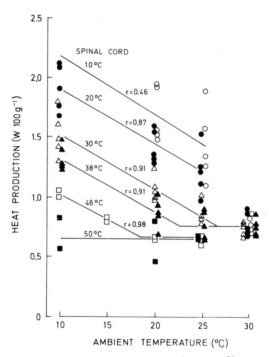

Fɪɢ. 9.2. The effect of spinal cord temperature on the rate of heat production in the rat at various ambient temperatures which are indicated on the abscissa. The results were obtained by perfusing the thermodes of five rats with water at 10 (○), 20 (●), 30 (△), and 38 °C (▲) and those of only two of them at 46 (□) and 50 °C (■). The temperature of the perfusate and the correlation coefficient are indicated close to each line. (From Banet and Hensel, 1976a.)

thus indicating a change in hypothalamic sensitivity. Peripheral warming decreased hypothalamic sensitivity for initiation of panting as well as for metabolic heat production, while peripheral cooling increased the sensitivity for both responses (Boulant and Gonzalez, 1977).

9.1.2 ANTERIOR HYPOTHALAMUS AND SPINAL CORD

In a number of investigations thermal stimulation of the spinal cord was combined with various thermal stimuli elsewhere in the body. Independent heating and cooling of the spinal cord and the skin in unanaesthetized pigs revealed a similar effect on oxygen consumption as found for hypothalamus and skin (Carlisle and Ingram, 1973b). Oxygen consumption increased with decreasing ambient temperature, and the threshold for this metabolic increase was shifted to higher ambient temperatures when the temperature over the spinal

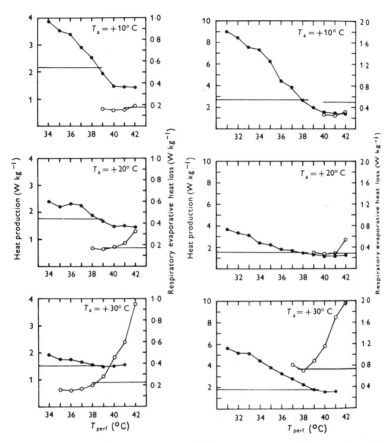

FIG. 9.3. Left: Effect of air temperature (T_a) on responses to thermal stimulation of hypothalamus and spinal cord in a conscious dog, when perfusion temperatures (T_{perf}) were of equal strength at these two sites. Heat production (M, filled circles, left ordinate) and respiratory evaporative heat loss (REHL, open circles, right ordinate) are plotted as functions of hypothalamus and spinal cord perfusion temperatures. Each point respresents the mean of nine single values. The horizontal lines show means of M and REHL during all control periods. Right: Effect of air temperature (T_a) on responses to thermal stimulation of hypothalamus and spinal cord, when perfusion temperatures (T_{perf}) were of equal strength at these two sites. (From Jessen, 1977.)

cord was lowered. A similar interaction of spinal and ambient temperatures was found in rats (Fig. 9.2) for metabolic responses to cooling (Banet and Hensel, 1976a). The data suggest an addition of spinal and cutaneous thermal signals.

Spinal and hypothalamic thermosensitivity, as assessed in conscious dogs by metabolic increase and evaporative heat loss, can be considered as equivalent (Jessen and Mayer, 1971; Jessen and Simon,

1971; Jessen, 1977; Mercer and Jessen, 1978b). The thermoregulatory responses will be enhanced when both sites are heated or cooled simultaneously, while they will be diminished, or even abolished, when the spinal cord is cooled and the hypothalamus is heated at the same time, or vice versa.

A quantitative evaluation of spinal cord and hypothalamic temperature combinations for equal rates of heat production and respiratory evaporative heat dissipation reveals an approximately additive interaction of both thermal drives (Jessen, 1977). Changes in ambient temperature altered the slope of the curves describing thermoregulatory heat production as a function of combined hypothalamic and spinal temperatures (Fig. 9.3). Heat production could be described as being approximately proportional to a product of linear drives determined by hypothalamus and spinal cord temperature on one hand and air temperature on the other hand. In contrast, respiratory evaporative heat dissipation was approximately proportional to the sum of drives determined by spinal cord, hypothalamus and air temperatures. The central threshold temperatures for heat production and respiratory evaporative heat dissipation were found to be differently affected by air temperature. This indicates that the mechanisms involved are to some extent functionally independent (Jessen, 1977).

9.1.3 HYPOTHALAMUS, SPINAL CORD AND "RESIDUAL INNER BODY"

In conscious goats hypothalamic and spinal cord temperatures were clamped by means of chronically implanted thermodes, and the "residual inner body" temperature was changed by use of an intravascular heat exchanger (Mercer and Jessen, 1978b). At an ambient temperature of 20 °C, hypothalamus and spinal cord were simultaneously cooled to 30 and 32 °C, respectively, which induced shivering and increased heat production. Consequently residual internal temperature, as monitored by aortic and rectal temperature, rose gradually to 40.5 °C. Then hypothalamic and spinal cord temperature were raised to approximately 40 °C and kept at this level for the following 3 h. Heat production returned to its control value of 1 W kg^{-1}. A short time later heat was removed from the body by means of the intravascular heat exchanger. The resulting nearly linear decline of residual internal temperature occurred first without any cold defence response. At 36 °C internal residual temperature cutaneous vasoconstriction appeared, while heat production still remained at the resting level. At 34.5 °C the shivering threshold was passed and heat production rose at a rate of $3\,W\,kg^{-1}\,°C^{-1}$, finally attaining a level of $4\,W\,kg^{-1}$ at a stable internal residual temperature of $33.5\,°C$. This rise in heat production occurred with hypothalamic and spinal cord temperatures being clamped at $40\,°C$, and must therefore be attributed to the effect of central or extracentral thermosensitive structures outside these areas. An effect of slight decreases in skin temperature due to vasoconstriction can also be ruled out, as has been shown by control experiments in which the skin was separately cooled.

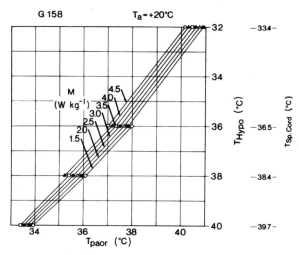

FIG. 9.4. Contour plots showing heat production (M) of a conscious dog as a function of residual internal temperature (T_{paor} = blood temperature) at various combined hypothalamic (T_{Hypo}) and spinal cord ($T_{Sp.\ Cord}$) clamp temperatures. The contour lines join combinations of equal heat production. (From Mercer and Jessen, 1978b.)

The residual internal temperature at which shivering occurred was directly related to the value of the combined hypothalamic and spinal cord clamp temperature. The higher hypothalamic and spinal clamp temperature were, the lower was the residual internal temperature at which heat production rose. Plots relating residual internal temperature to hypothalamic and spinal cord temperature at different levels of heat production show that the thermal input from the residual inner body is nearly of the same order of magnitude as that from hypothalamus and spinal cord (Fig. 9.4).

The only antagonistic responses to hypothalamic and spinal thermal stimulation were found for cardiac responses in anaesthetized and paralysed dogs (Göbel *et al.*, 1977). While hypothalamic heating decreased cardiac output and heart rate, spinal heating increased these parameters. Hypothalamic cooling had no effect on cardiac activity, but spinal cooling decreased both cardiac output and heart rate.

9.1.4 ANTERIOR AND POSTERIOR HYPOTHALAMUS

Independent temperature displacements of the anterior and posterior hypothalamus in conscious goats (Puschmann and Jessen, 1978) have shown that shivering and increased heat production can only be elicited from the anterior hypothalamus, whereas cooling the posterior hypothalamus resulted in an inhibition of shivering and heat production, and warming in an increase of these responses. Thus thermosensitivity of these allegedly integrative neurons affects shivering and heat production in a way inverse to the temperature sensing neurons in the anterior hypothalamus. It can be concluded that under natural conditions, where the temperature of anterior and posterior hypothalamus changes simultaneously, the thermoregulatory drives from both areas will counteract each other to a certain extent, and thus the overall thermoregulatory response may be smaller as when the POAH region is separately stimulated.

9.1.5 INTERACTION OF THERMAL DRIVES UNDER NATURAL CONDITIONS

Under natural conditions the thermal drives from various sites of the body may interact either in a synergistic or in an antagonistic way. On external cooling, cold receptors in the skin initiate and sustain processes of cold defence (shivering, non-shivering thermogenesis, vasoconstriction). These reactions are opposed by warmth-activated central thermoreceptive structures in hypothalamus, spinal cord, and perhaps other central and extracentral regions in the body core. This view is supported by the finding that external cooling of the paws in conscious cats leads to an increase in hypothalamic temperature (Kundt *et al.*, 1957b). Local warming of the preoptic-anterior hypothalamic (POAH) region suppresses metabolic reactions elicited by external cooling; warming the posterior hypothalamus has no such effect. This bridling effect of internal warm-sensitive structures can also be demonstrated by inactivating them through electrocoagulation. A steep increase in shivering (Andersson, 1970) or non-shivering thermogenesis (Wünnenberg and Brück, 1968a) appears at once.

Central cold-sensitive structures are only activated when the body becomes hypothermic after more extreme and prolonged cold exposure. Although cooling the POAH region induces vigorous metabolic reactions only if mean skin temperature is below the zone of thermal neutrality, there is some evidence that hypothalamic cold-sensitive structures may elicit some cold-defence reactions at normal skin temperature and even at normal body temperature. For example, small lesions in the medial preoptic region of cats produced a chronic significant decrease of 2 to 3 °C in mean colonic temperature and a

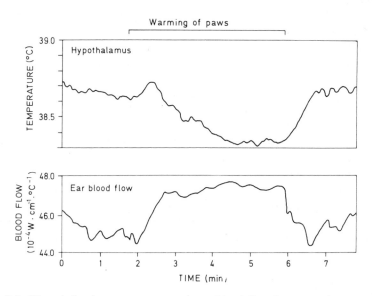

Fig. 9.5. Hypothalamic temperature and ear blood flow in a conscious cat when warming the paws. (From Kundt *et al.*, 1957b.)

lower rate of oxygen consumption at air temperatures between 19 and 27 °C (Squires and Jacobson, 1968). These observations were explained by the destruction of cold-sensitive structures concentrated within the medial preoptic region.

Heat dissipation mechanisms (panting, sweating, vasodilatation) can be activated by central heating of the body, e.g. during physical exercise, as well as by exposure to high ambient temperatures. Heating the paws in cats may lead to a transient fall in hypothalamic temperature (Fig. 9.5). At the same time, a marked reflex vasodilatation occurs (Kundt *et al.*, 1957b). This demonstrates that warming the skin can activate heat loss mechanisms at constant or even decreasing hypothalamic temperatures. It is well established that heat-dissipation mechanisms elicited by central warming can be inhibited by activation of peripheral cold receptors.

9.2 Combined Thermal Drives in Humans

In humans the separate and independent experimental manipulation of thermal drives from various sites of the body is more limited than in animal experiments. Another problem in humans is the measurement of temperatures inside the body. The temperature of the tympanic membrane has been suggested as an indicator of hypothalamic temperature and is now widely used for this purpose. There is still a discussion in progress about the correlation between both temperatures. Direct comparisons of both temperatures in cats and monkeys (Baker *et al.*, 1972) have revealed that the changes occurring in hypothalamic temperature during feeding, sleeping, and arousal are accompanied by similar changes in tympanic temperature, the latter being about 0.5 °C lower than the temperature of the hypothalamus. Tympanic temperature thus seems to be a relatively suitable indicator for hypothalamic temperature, provided that extreme changes in ambient temperature are avoided. Otherwise the influences from the skin of the head may cause considerable disturbances of tympanic temperature (Wurster, 1968; McCaffrey *et al.*, 1975a).

9.2.1 SWEATING RESPONSES

Sweat secretion in man has been shown to be controlled by both central and peripheral thermal influences (Robinson, 1949; Wyndham, 1966; Wyndham and Atkins, 1968; Stolwijk *et al.*, 1968; Benzinger, 1969, 1970, 1979; Wurster and McCook, 1969; Bullard *et al.*, 1970; McCook *et al.*, 1970; Timbal *et al.*, 1970, 1971; Nadel *et al.*, 1971; Houdas *et al.*, 1972, 1973; Ogawa, 1974; McCaffrey *et al.*, 1975b; Stolwijk and Hardy, 1977; Libert *et al.*, 1978, 1979; McCaffrey *et al.*, 1979). An example is shown in Fig. 9.6. Snellen (1972) described sweat secretion in man at rest and work as a function of the increase in average body temperature. Formally this finding also means that the internal temperature and the surface temperature act together in controlling the rate of sweat secretion.

   The question remains to be answered how far the effects of heating or cooling the skin on sweat secretion are due to neural impulses from the central nervous system or to local processes occurring in the skin itself. The same problem arises with regard to cutaneous blood flow. These mechanisms are discussed on p. 163.

By use of various methods it has been demonstrated that high skin temperatures can elicit reflex sweating responses and that thermal drives may thus originate from cutaneous warm receptors. When the skin temperature in the upper half of the body and the tympanic temperature were maintained at a constant level, the sweating rate in the upper part of the body was dependent on skin temperature of the

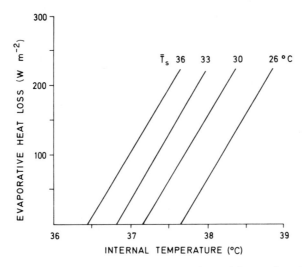

FIG. 9.6. Evaporative rate in man as a function of internal and mean skin temperatures. These lines are best fit of many experiments. (Modified from Nadel *et al.*, 1971.)

lower part between 34 and 40 °C (McCaffrey *et al.*, 1975b; McCaffrey *et al.*, 1979). This again shows the influence of skin temperatures above 33 °C.

There was also a dependence of sudomotor activity on the rate of change in skin temperature (Brebner and Kerslake, 1961; Colin and Houdas, 1968; Wurster and McCook, 1969; McCook *et al.*, 1970; Nadel *et al.*, 1971; Houdas *et al.*, 1972; Ogawa, 1974; Libert *et al.*, 1978, 1979). The higher the rate of change of average skin temperature, the lower was the static deviation of average skin temperature required for the onset of transient sweating in the forearm kept at a constant local temperature (Libert *et al.*, 1978, 1979).

9.2.2 METABOLIC RESPONSES

Various investigators (Stolwijk and Hardy, 1966, 1977; Cabanac, 1969; Nadel et al., 1970) have found a multiplicative relation between tympanic temperature (T_{ty}) and average skin temperature $(\bar{T}_s)$ for a given amount of cold-induced metabolic increase $(\varDelta M)$. According to Stolwijk and Hardy (1966), this relation can be described as follows:

$$\varDelta M = 70\,\mathrm{W}(36 - T_{ty})(34.1 - \bar{T}_s) \qquad (16)$$

Experiments with rapid intragastric cooling suggested the presence of a rate component of central temperature contributing to the regulatory response (Nadel et al., 1970). A question that cannot be answered concerns the site of central temperature reception in these experiments. Besides a thermal drive from the central nervous system, one must also consider the excitation of gastric thermoreceptors, the more so as cold receptors have recently been found in or near the stomach of cats (p. 115).

Rapid cooling of the skin, as it occurs during swimming in cold water, increases the metabolic rate without a considerable fall in oesophageal temperature (Nielsen, 1976). When oesophageal and skin temperatures are low and the skin is rapidly rewarmed during bicycling in warm air, the metabolic rate decreases to about the same level as observed with high oesophageal and skin temperatures. This dynamic effect of cutaneous temperature changes leads to a "hysteresis" of the metabolic response.

9.2.3 VASCULAR RESPONSES

Cutaneous vascular responses in humans (Benzinger et al., 1963; Benzinger, 1969; Wyss et al., 1974, 1975; Wenger et al., 1975a,b; Johnson and Park, 1979) were found to be proportional to core temperature at a given average skin temperature and also to depend on skin temperature. Figure 9.7 shows this relation for the cutaneous blood flow of the finger (Wenger et al., 1975a,b). The equation describing the additive interaction of oesophageal (T_{oe}) and average skin (T_s) temperature in the control of finger blood flow (BF) was

$$\mathrm{BF} = a_1 T_{oe} + a_2 T_s + b \qquad (17)$$

where a_1 and a_2 are constants, the proportion (a_1/a_2) of which was found to be 5.9 to 9.4 in various subjects. A similar relationship holds for the blood flow of the forearm (Wyss et al., 1974, 1975). It should be mentioned, however, that forearm blood flow is no reliable measure of skin blood flow and that during indirect heating skin blood flow in the forearm increases, whereas muscle blood flow decreases (Barcroft *et al.*, 1955; cf. Fig. 11.2).

When subjects were immersed in warm and cool baths in a certain sequence, thereby producing various combinations of skin and oesophageal temperatures, heat loss from the hand, evaporative heat loss and metabolic increase showed an approximately proportional change with oesophageal temperature for a given skin temperature

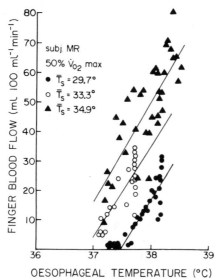

FIG. 9.7. Data from experiments in a human subject at three ambient temperatures, illustrating the effect of mean skin temperature ($\bar{T}_s$) on the relation between oesophageal temperature and finger blood flow, and showing the threshold for vasodilatation at low $\bar{T}_s$. (From Wenger *et al.*, 1975.)

(Cabanac and Massonet, 1977). In contrast to the findings shown in Fig. 9.7, the heat loss from the hand showed not only a shift in threshold but also a change of gain with increasing skin temperature. At 38 °C skin temperature, the response curve for heat loss as function of oesophageal temperature was 2.9 times as steep as it was at skin temperature of 28 °C.

9.3 Convergence of Thermal Inputs on Central Neurons

9.3.1 ANTERIOR HYPOTHALAMUS, SKIN AND SPINAL CORD

Multiple thermal and non-thermal influences affect the activities of hypothalamic neurons, and there is much evidence that neural and hormonal stimuli of various origin will modulate hypothalamic functions involved in autonomic and behavioural thermoregulation as well as feeding and drinking behaviour (for references see Hayward, 1977). Since thermoregulatory processes are influenced by local hypothalamic and peripheral temperatures, motor activity, level of arousal, hormonal secretion, and diurnal rhythm, the question can be raised whether these influences may change the activities of hypothalamic thermoregulatory neurons.

Current theories of central nervous control of body temperature suggest that information from temperature-sensitive neurons in the hypothalamus is combined in some way with that from other thermo-sensors in the central nervous system or elsewhere in the body. There is relatively little direct evidence for this view on the basis of single-unit recordings. Even if a neuron responds to more than one thermal input, this does not necessarily mean that integration occurs at this particular unit. The convergence of inputs may have occurred in other brain areas and then projected to the neuron whose activity is recorded (Nutik, 1971).

Wit and Wang (1968a) were able to show that hypothalamic neurons could be stimulated by heating the skin as well as by local heating of the hypothalamus. Hellon (1969, 1970a,b) found the same type of convergence in the anterior hypothalamus of rabbits and cats. About 75% of the thermosensitive cells responded to both local and peripheral stimulation, and they were usually affected in the same general direction by ambient and internal temperature. In some cases, there was also an opposite influence, in that cutaneous warming diminished the discharge of hypothalamic warm-sensitive units. The quantitative evaluation of the effects of central and peripheral inputs on single warm-sensitive neurons in the hypothalamus revealed various types of integration (Fig. 9.8). When the ambient temperature was changed, a parallel shift of the response curves to brain temperature was seen in several units, while others showed an increase or decrease in the slope of their sensitivity to local temperature (Hellon, 1972a,c).

The presence of hypothalamic neurons responsive to hypothalamic and skin temperatures was confirmed by various investigators in anaesthetized rabbits, rats and ground squirrels (Knox et al., 1973; Boulant and Bignall, 1973c; Boulant, 1974; Boulant and Hardy, 1974; Nakayama et al., 1979) as well as in unanaesthetized rabbits (Dymnikova et al., 1973; Reaves, 1977). For hypothalamic warm-sensitive neurons responding both to local and cutaneous heating, the average gain was found to be 11.2 (Reaves, 1977). Gain is defined by the sensitivity of the neuronal discharge to hypothalamic temperature divided by the sensitivity to ambient temperature.

Guieu and Hardy (1970, 1971) observed in anaesthetized rabbits that POAH neurons were influenced by thermal inputs from the spinal cord. In a systematic study the effects of hypothalamic, spinal and skin temperatures on the firing rate of POAH neurons were studied (Boulant and Hardy, 1974). Among the neurons with positive coefficients to preoptic temperature (warm-sensitive), about 60%

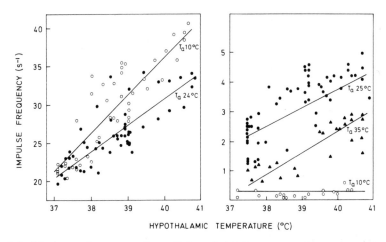

FIG. 9.8. Discharge frequency of two (left and right) warm-sensitive neurons in the rabbit hypothalamus at various hypothalamic and ambient (T_a) temperatures. (From Hellon, 1972c.)

responded also to spinal and cutaneous temperature, most of these with positive thermal coefficients. In addition, the incidence of extrahypothalamic thermosensitivity generally increased among the units having higher firing rates. Of the units with negative coefficients to preoptic temperature (cold-sensitive), about 73% responded also to extrahypothalamic temperature, about two-thirds of them with negative coefficients. There was no correlation between the incidence of extrahypothalamic thermosensitivity and the level of firing rate. Combining the results of this study with those of previous investigations, 76% of the units having dual thermosensitivities had similar coefficients for both hypothalamic and cutaneous temperatures. Neurons insensitive to preoptic temperature also responded very little to extrahypothalamic temperatures.

Figure 9.9 shows idealized response curves of POAH neurons sensitive to both hypothalamic and extrahypothalamic temperatures (Boulant and Hardy, 1974). Extrahypothalamic thermal stimulation not only changes the sensitivity to hypothalamic temperature but also alters the shape of the response curve. The same holds with respect to

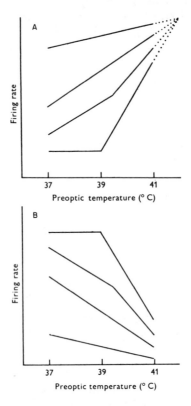

Fɪɢ. 9.9. Idealized thermoresponse curves of preoptic neurons in the rabbit showing how the sensitivity to preoptic temperature can change because of alterations in spinal or skin temperature: (a) refers to a unit with a positive preoptic temperature coefficient and (b) to a unit with negative coefficient. (From Boulant and Hardy, 1974.)

skin temperature. In the units having positive coefficients for preoptic temperature, an increased firing rate, due to extrahypothalamic temperature, generally resulted in a decreased preoptic thermosensitivity. In the units having negative coefficients for preoptic temperature, an increased firing rate, due to extrahypothalamic temperature, usually increased the preoptic thermosensitivity. This effect of firing rate on preoptic neuronal thermosensitivity was also found when a population of hypothalamic units was divided into groups based on their firing rates. As the level of firing rate increased in the warm-sensitive units, the preoptic thermosensitivity progressively decreased at temperatures above 39 °C, but generally increased at temperatures below 31 °C. In the cold-sensitive units, preoptic thermosensitivity generally increased as the firing rate at 38 °C increased (Boulant, 1974). Thus it is questionable whether a neuron, according to its type of response, can be classified as a primary thermosensor or an interneuron.

9.3.2 PREOPTIC AREA AND HIPPOCAMPUS

The hippocampus influences the firing rate of many neurons throughout the forebrain, including the preoptic region (Poletti et al., 1973; Hayward, 1977). In anaesthetized rabbits, preoptic and septal units were tested for their response to changes in preoptic temperature and electrical stimulation in the hippocampus, midbrain reticular formation, lateral lemniscus, and lateral spinothalamic tract (Boulant and Demieville, 1977). Brain stem stimulation affected 64% of the cold-sensitive neurons, 60% of the warm-sensitive neurons, and only 17% of the temperature-insensitive neurons. This corresponds well with the results of extrahypothalamic thermal stimulation (Boulant and Hardy, 1974). For the warm-sensitive neurons, brain stem stimulation had its greatest effect on those neurons with firing rates between 15 and $25 \, \text{s}^{-1}$. Hippocampal stimulation influenced a somewhat smaller percentage of thermosensitive neurons, but approximately in the same proportion. For the warm-sensitive neurons, the hippocampus had its predominant effect on neurons with a lower discharge rate between 10 and $15 \, \text{s}^{-1}$. These findings show that thermosensitive neurons in the preoptic and septal area can be characterized not only by their local thermosensitivity but also by their level of firing rate, their range of thermosensitivity and the type of synaptic input they receive from different neural sites.

9.3.3 POSTERIOR HYPOTHALAMUS

In the posterior hypothalamus (PH) of anaesthetized rabbits, about 9% of the neurons studied responded to thermal stimulation of the PH, the POAH region, and the spinal cord (Fig. 9.10), about 5% responded to the stimulation of two areas, and about the same number of units were sensitive only to PH temperature. No unit was observed which responded to spinal cord stimulation only. All units were located in the ventromedial area of the posterior hypothalamus (Wünnenberg and Hardy, 1972). These findings were generally confirmed in conscious rabbits (Dymnikova, 1973). About half of the PH neurons responding to POAH stimulation increased their frequency with POAH heating, and the remainder increased its frequency on POAH cooling.

Similar results were found in anaesthetized cats (Nutik, 1973a,b), where about 9% of the units in the posterior hypothalamus responded

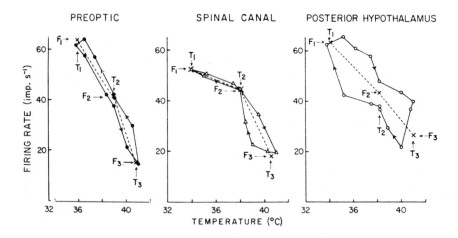

FIG. 9.10. Firing rates of a unit in the rabbit posterior hypothalamus as function of temperature change at various sites. Solid lines, with arrows indicating temporal course, connect data points; dashed lines connect average values. (From Wünnenberg and Hardy, 1972.)

to POAH temperature and about 5% to skin temperature, approximately two-thirds of the units increasing their firing rate with cooling of both sites. As found already by Wünnenberg and Hardy (1972), some of the temperature-sensitive PH units showed hysteresis when dynamic thermal stimuli were applied, while other units responded practically without hysteresis.

9.3.4 MIDBRAIN AND MEDULLA OBLONGATA
In contrast to the thermosensitive neurons in the posterior hypothalamus, numerous units in the midbrain reticular formation (MRF) in

anaesthetized rabbits were sensitive to local cooling and to cooling of the skin, but did not respond to thermal stimuli applied to the POAH region (Cabanac and Hardy, 1969; Nakayama and Hardy, 1969). Most of the MRF neurons responding to cutaneous cooling were of the non-linear type. In the same species, about 18% of the locally cold-sensitive MRF neurons decreased their firing rate with increasing temperature of the spinal cord (Fig. 9.11), and about 8% of the units increased their rate, while about the same proportions of locally warm-sensitive units responded in the opposite way to spinal heating than did the cold-sensitive neurons (Hori and Harada, 1976).

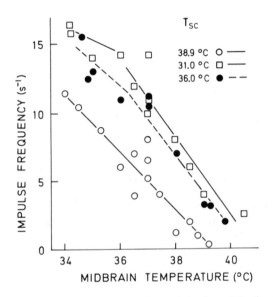

FIG. 9.11. Response curve of a cold-responsive unit in the midbrain of the rabbit as function of mid-brain temperature at three different levels of spinal cord temperature (T_{sc}). (From Hori and Harada, 1976.)

In the reticular formation of the medulla oblongata, about 50% of the locally thermosensitive neurons responded positively or negatively to changes in skin temperature (Inoue and Murakami, 1976).

9.3.5 SPINAL CORD

In the spinal cord of anaesthetized cats, practically all of the neurons sensitive to local cooling or warming are also influenced by temperature changes in the skin (Simon, 1972). Skin cooling below 38 °C led to an increase in activity of the spinal cold-sensitive units (Fig. 9.12), while the spinal warm-sensitive units were only little influenced by changes in skin temperature.

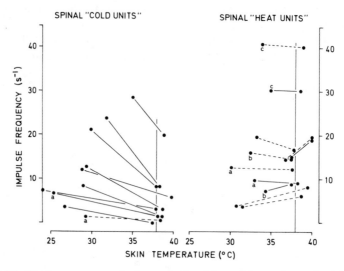

FIG. 9.12. Discharge rates of nine spinal cold-sensitive units and eight spinal heat-sensitive units in the cat as influenced by changes of skin temperature in the range below 40 °C. The dots connected by straight lines demonstrate the average responses of one or more experimental periods performed in one unit at constant vertebral canal temperature below 40 °C (solid lines) and above 40 °C (interrupted lines); identical letters indicate results obtained from the same unit. (From Simon, 1972.)

10
Central Thermoregulatory Connections

10.1 Sites of Integration

10.1.1 HYPOTHALAMUS

An essential role in the processing of thermal signals is ascribed to the posterior hypothalamus. This view can be supported by results of single-unit recordings which have shown that integration of thermoreceptive afferents from skin, spinal cord and POAH occurs in neurons in the posterior hypothalamus (p. 135). It is unlikely that the posterior hypothalamus acts as primary thermosensor for autonomic temperature regulation, because the density of thermosensitive neurons is relatively low and, as shown by experiments in squirrel monkeys, the autonomic thermoregulatory responses elicited by thermal stimulation of this region are either weak (Fig. 7.3) or even opposed to those elicited from the POAH region (cf. p. 126).

The perpetuation of patterns from the preoptic region through the caudal posterior hypothalamus gives support to the idea expressed by Birzis and Hemingway (1957) that some efferent pathways involved in cold shivering are organized in the POAH and septal regions and pass caudally to the ventral posterior hypothalamus. The present data also support the proposal (Wünnenberg and Brück, 1970; Simon, 1974) that the thermosensitive impulses originating in the spinal cord reach the posterior hypothalamus in the area through which shivering pathways pass.

Using the method of selective microcuts, Lipton *et al.* (1974) found that rats in which the POAH region was disconnected from the neighbouring tissue showed disturbances in regulation against both heat and cold. Even the separation of the preoptic region from the anterior hypothalamus caused thermoregulatory deficits either at high or at low ambient temperatures. In contrast, medial cuts rostral and caudal to the POAH region had no effect on thermoregulation, but parasagittal cuts impaired mainly the regulation against cold. It was therefore concluded that the pathways for cold-defence reactions pass laterally from the POAH region, presumably in the medial forebrain bundle. In addition, lesions in the medulla oblongata also impaired thermoregulation.

Gilbert and Blatteis (1977), using similar techniques, not only studied the changes in body temperature but also differentiated the control actions. In rats, microcuts which separated the preoptic region as well as the connections between the POAH region and the median forebrain bundle prevented cold-induced cutaneous vasoconstriction but did not impair shivering activity and heat-induced cutaneous vasodilatation. These results suggest that different sites in the hypothalamus may separately control cold-induced vasoconstriction and shivering activity as well as heat-induced vasodilatation.

It was assumed that the inputs from cutaneous cold receptors may be diverging toward two separate sites in the hypothalamus (Fig. 10.1). One site controlling thermoregulatory cutaneous vasoconstriction may be located in the preoptic area and connected to the medullary pressor area via the medial forebrain bundle. Another site controlling shivering activity may be located in the posterior hypothalamus and connected to an additional pathway from cutaneous cold receptors.

Since heat-induced cutaneous vasodilatation was not impaired by the cuts which prevented cold-induced cutaneous vasoconstriction, the conclusion seems justified that heat-induced vasodilatation may be controlled separately from cutaneous vasoconstriction at a site more caudal than the preoptic area, possibly in the medulla (Fig. 10.1). There is also some evidence for a local heat sensitivity of this area (p. 101).

It appears that the entire brain stem from the septal-preoptic region through the caudal posterior hypothalamus participates both in the sensing of temperature, the integration of afferent pathways from sites of peripheral thermal stimulation, and the organizing of the efferent pathways. However, little is known about the integration of thermoregulatory pathways. Thus it is still a matter of conjecture whether the processes of heat production and heat dissipation are correlated or separated. Cabanac (1975) has summarized the evidence supporting the conclusion that the various functions involved in regulation against heat and cold are largely independent of one another. It may even be possible that any of the autonomic control actions, such as

shivering, vasomotion, sweating, as well as thermoregulatory behaviour are acting through more or less independent networks.

Although in many cases thermoregulatory responses seem to be
interchangeable and thus appear as though they were activated by
one and the same thermal input, there are indications for a relative
independence of various loops: (1) The thresholds for heat production

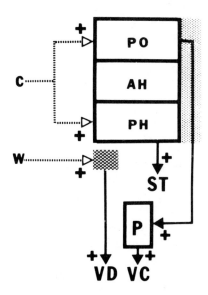

FIG. 10.1. Schematic representation of postulated intrahypothalamic thermoregulatory pathways in the rat. C, Cutaneous cold stimulus; W, cutaneous warm stimulus;
PO, preoptic area; AH, anterior hypothalamus; PH, posterior hypothalamus; ST,
shivering thermogenesis; P, pressor area of medulla; VD, cutaneous vasodilatation;
VC, cutaneous vasoconstriction; dotted area = medial forebrain bundle; hatched
area = vasodilator motor area; dashed lines = afferent pathways; solid
lines = efferent pathways; + = stimulatory. (From Gilbert and Blatteis, 1977.)

and heat dissipation can be varied separately and independently, e.g.
by pharmacological agents (Cabanac, 1975) or by long-term adaptation (Chapter 15). (2) Hypothalamic lesions can impair one of the
thermoregulatory responses and leave others intact (Gilbert and
Blatteis, 1977). (3) Thermal stimulation or lesions at various sites of
the central nervous system can be followed by a dissociation between
autonomic thermoregulation and thermoregulatory behaviour (Chapter 13).

Histochemical (Lindvall and Björklund, 1974; Kobayashi *et al.*, 1975; Roizen, 1976; Merker *et al.*, 1977) and physiological (Zeisberger and Brück, 1971a, 1976; Szelényi *et al.*, 1976, 1977; Zeisberger *et al.*, 1977) investigations in guinea pigs and rats have revealed a noradrenergic system ascending from the brain stem reticular formation to the hypothalamus. This pathway could be activated by electrical stimulation of the lower brain stem, leading to a rise in oxygen uptake and an increase in body temperature. Additional micro-injection of an adrenergic alpha receptor blocker, phentolamine, into the hypothalamus blocked this effect completely (Szelényi *et al.*, 1977). This pathway is assumed to be connected with hypothalamic integrative neurons (cf. p. 232); it is not identical with the pathway from peripheral thermoreceptors to the hypothalamus (Brück, 1978c).

10.1.2 MEDULLA OBLONGATA

While the hypothalamus may be the principal site of thermoregulation, some independent but less powerful thermoregulatory structures exist in the medulla oblongata (Chai and Lin, 1972, 1973; Lipton, 1973; Lipton *et al.*, 1974). On the one hand, thermoregulatory responses can still be elicited in rats by local thermal stimulation of the medulla oblongata when lesions are made in the POAH region (Lipton, 1973), or in rabbits after decerebration (Chai and Lin, 1973). On the other hand, lesions in the medulla oblongata in rats at sites known to be thermosensitive (p. 101) produce deficits in resistance to either overheating and chilling (Lipton *et al.*, 1974). Furthermore, as concluded from the results of microcuts in the hypothalamic region, neural structures involved in the control of vasodilatation have been postulated in the medulla oblongata (Gilbert and Blatteis, 1977). However, no detailed evidence is available as to how thermosensitive or thermointegrative structures in the medulla are connected to hypothalamic or afferent thermal pathways.

10.1.3 SPINAL CORD

Studies in spinalized animals have shown that some integration of thermal afferents may already occur at the spinal level. In spinalized, unanaesthetized rabbits, shivering could be produced by cooling the vertebral canal (Kosaka and Simon, 1968a,b; Thauer, 1970; Simon, 1974). In addition, the observations in unanaesthetized animals proved a marked shift in shivering threshold after spinal transection.

Although in intact animals the increase in activity began after a mean reduction of the peridural temperature to 37.1 °C, a decrease of the same temperature to a mean value of 35.3 °C was necessary in spinalized rabbits. The analysis of the electromyogram showed only minor differences between normal and spinal animals. Thus the experiments not only confirm the existence of thermosensitive structures in the spinal region, but also prove that the spinal cord, at least in these animals, is equipped with mechanisms for the production of shivering.

Paraplegic patients with complete high spinal transection (T3) are able to sweat on the lower part of the body when placed in a 60 °C room (McCook et al., 1970). This response was suggested to be due to a spinal reflex elicited by afferents from cutaneous warm receptors. However, in quadriplegic patients with high cervical transections (C8 to C3), a marked anhidrosis and a subsequent tendency toward hyperthermia in warm environments was found (Totel et al., 1971). In any case, sudomotor and vasomotor thermoregulatory responses in paraplegic patients are severely impaired (Normell, 1974; Downey et al., 1976).

No evidence for a spinal reflex leading to metabolic increase during cooling of the skin was found in spinal man. When the insentient portions of paraplegic patients with spinal transections at T2 to T11 were cooled, an increase in oxygen consumption occurred only when the ear canal temperature fell to about 35.7 °C, indicating that deep body thermosensors were the origin of the response (Downey et al., 1969, 1971). On the basis of available evidence the conclusion seems justified that the human spinal cord has no essential function as an integrative centre for autonomic thermoregulatory functions.

10.2 Neuronal models

Besides the formal descriptions of some input/output relations of the thermoregulatory system in terms of cybernetics, several authors have tried to develop physiological models that interpret the input/output relations in terms of neuronal networks and specific properties of the neurons based on single unit recordings and synaptic interference studies. However, even these models are highly abstractive and by no means a description of real neural connections, despite the fact that in some graphs the ganglion cells are drawn in a more or less realistic manner. A neuronal model may be "just another analogue model, built of neuronal units in place of mechanical devices" (Bligh, 1973). However, such models may bring some order into the experimental

findings and stimulate further progress towards an understanding of
thermoregulation. A collection of neuronal models proposed by
various authors is given by Bligh (1973, 1974, 1979).

Figure 10.2 shows some principal possibilities of neuronal input/
output relations in the thermoregulatory system. The model in Fig.
10.2A has independent networks for heat production (HP) and heat
dissipation (HD). The pathways for heat production are activated by

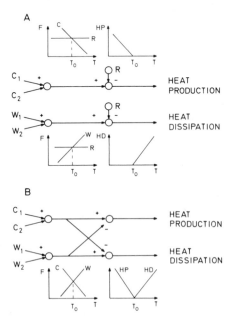

FIG. 10.2. Simplified neuronal models of temperature regulation. A: Separate systems
for heat production and dissipation. B: Systems with reciprocal inhibition. C, Cold
receptor; W, warm receptor; R, reference; HP, heat production; HD, heat dissipation;
F, frequency of thermoreceptor; T, temperature; T_o, set point. For further explanation
see text.

cold receptors (C), and the pathways for heat dissipation are activated by warm receptors (W). The model includes thermally insensitive reference neurons (R) which generate a set point (T_0) by
comparison between the frequency (F) of thermoreceptors and that of
the reference units. Separate set points are postulated for heat
production and heat dissipation.

Another model (Fig. 10.2B) was originally proposed by Vendrik
(1959). The set point is presumed to be generated by a comparison

between the frequencies of simultaneously active cold and warm receptors. In order to generate threshold functions for heat production and heat dissipation which normally do not overlap, a reciprocal inhibition of both channels is introduced (Bligh, 1979). Without additional assumptions this model hardly accounts for the fact that the thresholds for heat production and heat dissipation can be shifted separately and independently.

A neuronal model for the interaction of cold-induced shivering and non-shivering thermogenesis in neonatal temperature regulation (Brück, 1978a,b) is discussed in Chapter 16.

10.3 Central Neurotransmitters

Starting from investigations by von Euler *et al.* (1943) and Feldberg and Myers (1964), evidence for the view that certain neurotransmitters are involved in temperature regulation has been derived from various approaches which have been refined during recent years and have largely been applied in unanaesthetized animals. In the experiments hitherto performed, only a few systematic studies of the quantitative relations between input from thermosensors and output to thermoregulatory effectors have been made in connection with neural transmitters (e.g. Zeisberger and Brück, 1971a, 1976; Szelényi *et al.*, 1976, 1977). A rise or fall in core temperature does not suffice as an indicator for the changes that have occurred in the thermoregulatory system.

In the following, experiments in unanaesthetized cats and primates will mainly be discussed. These animals give reasonably similar body temperature responses to intraventricular injection and microinjection of drugs. Quite different responses have been reported in other species such as sheep, rabbits and rats (for references see Hellon, 1975; Bligh, 1975, 1979; Satinoff, 1979). Although there is no direct evidence as yet, the probability seems relatively high that analogies for humans may be derived from primates which have been proved to possess thermoregulatory mechanisms very similar to that of humans. Some findings in guinea pigs will be dealt with in connection with long-term adaptation (Chapter 15) and neonate temperature regulation (Chapter 16).

10.3.1 MONOAMINES AND TEMPERATURE REGULATION

In order to ascertain the significance of injected chemical substances for thermoregulation, one must select the appropriate parameters to measure and must give the injections at an exact site relevant to thermoregulation. Therefore, the application of microinjections into precisely located areas of the brain as compared to intraventricular injections was a considerable step forward.

Table 10.I shows some results of microinjections and intraventricular injections of various substances in unanaesthetized cats. Microinjections of noradrenaline (NA) into the POAH region led to a consistent fall in body temperature. The site of action was located just below the anterior commissure in a plane (AP 14 to 16; Jasper and Ajmone-Marsan, 1954) which joints the rostral borders of the anterior commissure and the optic chiasm (Cooper et al., 1976; Metcalf and Myers, 1978).

Thermoregulatory behaviour following microinjections of NA in cats was described by Cooper et al. (1976). At ambient temperatures of 10 °C, the cats usually sat quietly in a huddled position. After NA microinjections (2.5 to 20 μg) they frequently walked around and stretched, but none of the animals adopted the open posture seen at higher temperatures. Ear temperature and respiratory rate increased, but open mouth panting was never observed. At 35 °C ambient temperature, the animals generally adopted an open, sprawled out posture. Following NA microinjection (2.5 to 10 μg) ears became hot and flushed, and respiratory rates increased dramatically up to 15-fold in some cats. At both ambient temperatures there was a dose-dependent fall in body temperature up to 1 °C in the warm environment and 3 °C in the cold environment.

Numerous observations of thermoregulatory behaviour following intraventricular or microinjection of noradrenaline, 5-hydroxytryptamine and acetylcholine have been performed in rats (for references see Satinoff, 1979). The results seem to be relatively inhomogenous and will not be discussed here.

Attempts have been made to test whether the responses to exogenous amines are the same as those evoked by the animal's endogenous supplies. This can be done by depleting the presynaptic stores, by interfering with the mechanisms that inactivate the released transmitter, or by blocking transmitter action on postsynaptic membranes. NA is mainly inactivated by reuptake into the presynaptic endings, and this process can be inhibited by imipramine and N-desmethylimipramine (Cranston et al., 1972). After intraventricular injection of these drugs in cats, the body temperature decreased, whereas in rabbits an increase occurred. Furthermore, these temperature changes were diminished if the endogenous stores of NA had previously been depleted by α-methyl-p-tyrosine (AMPT).

... hypothalamic microinjections of biogenic amines in unanaesthetized cats

Drug	Technique	Ambient temperature or state of animal	Effect on control actions	Effect on body temperature	Remarks	References
NA[a]	Microinjection POAH	20–22 °C	Vasodilatation No panting Shivering stopped	1–2 °C decrease	Dose-dependent response	Feldberg and Myers (1965) Rudy and Wolf (1971)
NA	Microinjection into posterior or ventromedial hypothalamus	normal	None	None		Rudy and Wolf (1971)
NA	Microinjection POAH	10, 20, 35 °C	At higher ambients heat loss activated, at lower ambients heat production inhibited	0.5–3 °C decrease	Dose-dependent response	Cooper *et al.* (1976)
NA	Microinjection POAH	20–22 °C	Vasodilatation sometimes tachypnoea	0.5–2 °C decrease	Dose-dependent response, antagonized by phentolamine	Metcalf and Myers (1978) Ruwe and Myers (1978)
DA	Microinjection POAH		Vasodilatation slight tachypnoea	0.5–1 °C decrease	Dose-dependent response, antagonized by haloperidol and D-butaclamol	Ruwe and Myers (1978)
5-HT	Microinjection into hypothalamus or ICV	20–22 °C	Vasconstriction, shivering or vasodilatation, tachypnoea	increase or decrease or nothing		Feldberg and Myers (1964, 1965) Wendlandt (1972)

[a] Abbreviations: NA, noradrenaline; DA, dopamine; 5-HT, 5-hydroxytryptamine; ICV, intracerebro-ventricular; POAH, preoptic and anterior hypothalamic.

If a substance is considered to be a neurotransmitter, its release in the hypothalamic tissue has to be shown under thermoregulatory conditions where the synapses are likely to be activated. Myers and Sharpe (1968) devised a system in which bilateral push–pull cannulas were implanted into the anterior hypothalamus. With tracer techniques it was found that the release of NA was greatly increased in cats exposed to ambient temperatures of 40 °C but not of 10 °C (Myers and Chinn, 1973). The sites of this release were only in the anterior hypothalamus between the anterior commissure and the optic chiasma and within 1.5 mm of the midline.

Responses to 5-hydroxytryptamine (5-HT) in cats have not been as consistent as those to NA (Table 10.I). Various authors report hyperthermic, isothermic or hypothermic effects without a satisfactory explanation for these differences. There is evidence for a release of 5-HT in perfusates of the third ventricle (Feldberg and Myers, 1966), but this release has not been tested with thermal stimuli as has been done in the case of monkeys (p. 150). There is still uncertainty about the action of 5-HT in the cat's thermoregulatory system.

Acetylcholine (ACh) is an obvious transmitter candidate in hypothalamic pathways although it does not seem to be present in particularly high concentrations. The evidence summarized by Hellon (1975) indicates the presence of "nicotinic" receptors in heat dissipation pathways and "muscarinic" receptors in heat conservation pathways. Microinjection experiments (Rudy and Wolf, 1972) have shown that cholinoceptive sites are widely scattered between the levels of the optic chiasma and the mamillary bodies, in contrast to the restricted sites in the POAH area for NA and 5-HT.

When the metabolic precursor of NA, dopamine (DA), was given intraventricularly or by microinjection into the POAH region of conscious cats, a dose dependent decrease in rectal temperature occurred (Kennedy and Burks, 1974; Ruwe and Myers, 1978). This response was somewhat depressed by the alpha-receptor blocking agent phentolamine, but completely abolished by the dopamine antagonists haloperidol and D-butaclamol. The beta-receptor blocking agent practolol was ineffective. DA does not seem to act through release of endogenous NA, because the administration of 6-hydroxytryptamine (6-HDA), a neurotoxin that causes degeneration of noradrenergic nerve endings, did not prevent the hypothermic responses to DA, while under these conditions the action of tyramine, an indirectly acting sympathiocomimetic amine, was blocked. Therefore it was assumed that specific dopamine receptors are involved. This view was further supported by the observation that in some cases microinjections of DA into the POAH were effective at sites other than those responding to NA. In addition, environmental warming to 35 to 45 °C released DA or NA at circumscribed sites of the POAH region. However, the physiological part played by DA is not yet established.

Rhesus monkeys respond to intraventricular injections of NA and 5-HT in the same manner as cats and dogs (Feldberg *et al.*, 1967). A

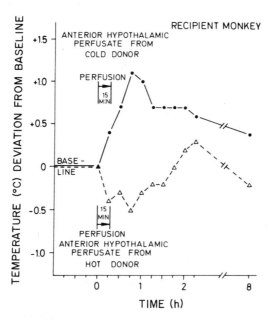

FIG. 10.3 Deviation of body temperature of a monkey (recipient). Black arrows indicate the perfusion interval (15 min). In the first experiment (●) the recipient's anterior hypothalamus was perfused with perfusate from a corresponding site in a cooled donor monkey. In the second experiment (○), 3 days later, perfusate from a heated donor was transfused to the same site. (From Myers and Sharpe, 1968.)

more precise localization of the active sites for NA, 5-HT and ACh was possible by a systematic exploration of the hypothalamus and brain stem in conscious monkeys with microinjections of these substances (Myers and Yaksh, 1969) in volumes of 1 μl or less. Nearly all sites from which the hyperthermic effect of 5-HT and the hypothermic effect of NA were observed were closely grouped together in the anterior part of the hypothalamus, lying between the anterior commissure and the optic chiasma. In contrast to this grouping, the points from which ACh was effective were scattered fairly uniformly throughout the hypothalamus. Usually a temperature rise was observed but, in a few cases, a decrease was caused when ACh was injected into the posterior hypothalamus.

Using push–pull cannulas chronically implanted into the anterior hypothalamus, Myers and Sharpe (1968) exposed a donor monkey to hot and cold environments and microinjected the perfusate fluid from the hypothalamus of the exposed monkey into the anterior hypothalamus of the donor monkey (Fig. 10.3). The perfusate from the cold donor induced a temperature rise in the recipient, while the perfusate from the warm donor caused a fall in temperature. Bioassays of push–pull perfusions for 5-HT (Myers and Beleslin, 1971) and ACh (Myers and Waller, 1973) have shown that 5-HT was released only during external cooling and only from sites in the POAH. These sites and the sites of sensitivity for microinjections of exogenous 5-HT coincided. More complex results were obtained from the assay of ACh during similar experiments. In the POAH region, ACh release tended to be enhanced by external cooling and suppressed by warming. This result corresponds to the hyperthermic action of ACh when administered by microinjection into the POAH region. More caudal microinjections produced mixed results which corresponded to some extent with the hyper- and hypothermic actions of exogenous ACh. There have been no comparable assays for NA release in the monkey.

As shown by microassays in monkeys, ACh was released during 5-HT hyperthermia from loci in both thalamus and mesencephalon, the effect being maximal when 5-HT was injected into the rostral hypothalamus (Myers and Waller, 1975).

The neurotoxins 5,6-dihydroxytryptamine(5,6-DHT) and 5,7-dihydroxytryptamine(5,7-DHT) produce degenerative changes in the presynaptic elements of serotonergic neurons (Dali *et al.*, 1974). A powerful and long-term reduction of 5-HT can be achieved by these neurochemical tools. In the monkey, microinjections of 5,6-DHT into serotonin-sensitive sites caused a severe chronic impairment of thermoregulation in cold environments (Fig. 10.4) and, with high doses, also, a certain impairment of regulation in warm environments (Waller *et al.*, 1976; Myers, 1978). However, exogenous 5-hydroxytryptamine still evoked a temperature response nearly identical to that seen before 5,6-DHT treatment.

No direct evidence for the significance of monoamines in human thermoregulation is as yet available. Phenylketonuric children, who would be expected to have a low level of brain 5-HT and thus a disturbed cold defence, showed just the reverse (Blatteis *et al.*, 1973).

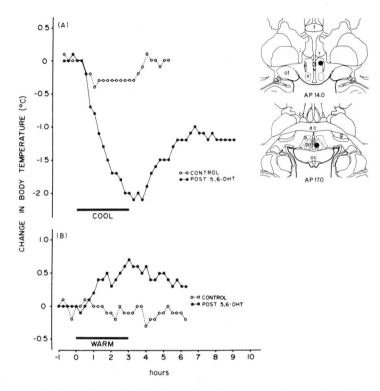

FIG. 10.4. Deviation from the baseline body temperature of a monkey during a 3 h period of cooling (4 °C) one day (A) after the microinjection of 3.0 μg 5,6-DHT and during a similar period of warming (40 °C) two days (B) after these injections. The injections were made simultaneously at the sites denoted in the histological inset. (From Waller *et al.*, 1976.)

Compared with normal children, those with phenylketonuria had undisturbed responses to cold, but in the heat their sweating was diminished. However, it is dubious whether this finding can be related to brain amine levels.

10.3.2 RESPONSE OF THERMOSENSITIVE NEURONS

In order to test the responses of thermosensitive neurons to locally applied substances, multibarrel microelectrodes were used which allowed simultaneous recording from single units and micro-iontophoretic application of drugs. A great deal of the results was not as clear-cut as could be expected from the microinjection experiments. In cats and rats, Beckman and Eisenman (1970) applied NA, 5-HT

and ACh by microiontophoresis to preoptic single neurons whose
response to changes in local temperature had been tested previously.
One type of warm-sensitive cells responded to temperature rises and
to the iontophoresing current but apparently not to the substances.
These neurons may have no synaptic input and could be primary
thermoreceptive units. Other groups were classed as warm-sensitive
and cold-sensitive interneurons; they responded to the drugs but not
to the current. In rats and cats, the warm-sensitive neurons of the
latter group were accelerated by ACh and slowed by NA, while the
firing rate of cold-sensitive neurons was accelerated by NA. This
would correspond to a certain extent to the thermoregulatory responses in rats when the substances are administered by microinjection into the anterior hypothalamus (for references see Hellon, 1975),
but would be in contradiction to the findings in cats where microinjections of NA into the POAH region act like an excitation of warm-
sensitive structures and an inhibition of cold-sensitive ones (Table
10.I). Other investigators (Jell, 1973; Ford, 1974) came to different
results but arrived essentially at the same conclusion, that no
clear-cut correlation exists between thermal response pattern and
drug sensitivity. The activity of preoptic warm-sensitive and temperature-insensitive neurons in dogs was depressed by intraventricular
administration of both 5-HT and NA (Cunningham *et al.*, 1967)
which, in case of NA, is in contradiction to the effect expected from
the thermoregulatory responses to microinjections.

The only satisfactory correlation so far between the effect of
biogenic amines on warm- and cold-sensitive hypothalamic neurons
on the one hand and thermoregulatory responses to microinjections
on the other hand was found in rabbits. In the investigations by
Murakami (1973), about one-third of the temperature-sensitive
neurons in the POAH region did not respond to any drugs; the
remaining warm-sensitive neurons were excited by 5-HT and depressed by NA, whereas the opposite was seen in cold-sensitive neurons.
Excitatory as well as inhibitory effects were seen both in warm- and
cold-sensitive neurons following the application of ACh. Hori and
Nakayama (1973) found that the firing rate of nearly all warm-
responsive neurons in the POAH region was consistently increased by
microiontophoresis of 5-HT and decreased by NA, while the opposite
effect was seen in cold-responsive neurons (Table 10.II). No response
was seen when ACh was given.

Thermosensitive neurons in the midbrain reticular formation
reacted in the opposite way as did the hypothalamic neurons, in that
the cold-sensitive neurons increased firing with 5-HT and decreased it

TABLE 10.II

Effects of biogenic amines on central thermoresponsive neurons in rabbits

Region	Thermal response	Substance	Units tested n	Frequency increase n(%)	Frequency decrease n(%)
POAH	None	5-HT	54	15	59
		NA	51	27	29
		ACh	44	23	2
	Warm	5-HT	17	88	0
		NA	17	0	76
		ACh	17	0	0
	Cold	5-HT	7	0	86
		NA	6	83	0
		ACh	7	0	0
Midbrain reticular formation	None	5-HT	32	6	25
		NA	32	16	19
		ACh	31	19	13
	Warm	5-HT	5	0	100
		NA	5	0	0
		ACh	5	0	20
	Cold	5-HT	14	100	0
		NA	14	0	57
		ACh	14	43	0

Data from Hori and Nakayama (1973).

with NA, whereas the warm-sensitive units decreased their firing rate on application of 5-HT but did not respond to NA. ACh increased the firing rate of about half of the cold-sensitive neurons but was ineffective on the warm-sensitive units.

These findings are in accordance with thermoregulatory responses in rabbits (for references see Hellon, 1975), in particular with the results obtained by microinjection of drugs into the POAH region of conscious animals (Cooper *et al.*, 1965), where 5-HT induced hypothermia, NA induced hyperthermia, and ACh was ineffective. The relatively uniform results in rabbits and the rather conflicting findings in cats and dogs remain still to be explained.

Summing up the results of all these approaches, Hellon (1975) comes to the conclusion that the rather inhomogenous and partially conflicting results may best be explained by the hypothesis that monoamines serve as neurotransmitters in afferent pathways which relay information from thermosensors in the skin and elsewhere to the hypothalamus, while ACh acts in integrating pathways in the hypothalamus. Evidence for this view is derived from the following facts: (1) NA and 5-HT are confined almost entirely to hypothalamic nerve endings, the cell bodies of which are found more caudally in the raphe nuclei for 5-HT and in ventral areas of the brainstem for NA. (2) Electrical stimulation of these cell body areas excites POAH neurons which are sensitive to hypothalamic temperature (Eisenman, 1974a). Thus there is a specific link between the monoaminergic neurons and more rostral neurons presumably involved in the controlling system. (3) External heating can excite the firing rate of neurons in the midbrain raphe nuclei (Weiss and Aghajanian, 1971; Jahns, 1976, 1977). In addition one of the inputs of these nuclei comes from the anterolateral spinal tract which carries afferent thermal fibres (Brodal *et al.*, 1960). (4) External heating and cooling can also cause the release of NA and 5-HT respectively from localized sites in the POAH (Myers and Beleslin, 1971; Myers and Chinn, 1973; Myers and Waller, 1973).

However, the finding remains that in the rabbit nearly all neurons sensitive to local POAH temperatures also respond to microiontophoresis of either NA or 5-HT. This seems neither compatible with the assumption that NA and 5-HT are acting only on afferent pathways nor with the fact that virtually no serotonergic neurons have been found in the hypothalamus.

11
Control of Effector Systems

In connection with thermoreception and its significance for temperature regulation, only a general outline will be given of the control of thermoregulatory effector systems. For more detailed information a number of books and reviews may be consulted (Hardy *et al.*, 1970; Bligh, 1973, 1976; Hensel, 1973b; Hensel *et al.*, 1973; Cabanac, 1975; Ingram and Mount, 1975; Le Blanc, 1975; Houdas and Guieu, 1978; Brück, 1978a,b). Some emphasis is put on peripherally induced changes in effector mechanisms because local influences may interfere with those elicited by the activity of internal and external thermosensors.

11.1 Centrally Induced Processes

The short-term control actions are initiated via nervous pathways; hormonal factors play a role only in long-term alterations of the thermoregulatory system. Two nervous pathways take part in this transmission: (i) the spinal and supraspinal motor system and (ii) the sympathetic nervous system.

11.1.1 SHIVERING

The descending tracts of the "shivering pathway" leave the posterior hypothalamus and run caudally through the midbrain tegmentum and the pons, close to the rubrospinal tracts. Shivering is controlled via the motor system whose supraspinal pathways (tractus cerebrospinalis and reticulospinalis) and peripheral portions (α- and γ-motoneurons) are well known.

Cooling the spinal cord leads to an increased excitability of the motoneurons and thus to shivering (Klussmann and Pierau, 1972). Thus the spinal cord is not only a site of thermosensitive structures

that transmit impulses to the posterior hypothalamus via ascending
pathways but it also contains thermosensitive motoneurons that can
reinforce the effector mechanisms of the system regulating against
cooling. This spinal mechanism is likely to be activated during
prolonged cooling, when not only the peripheral but also the central
temperature decreases. In this respect more information about the
temperature course in the vertebral canal under natural conditions
would be required.

Golenhofen (1963, 1970) has shown that in human subjects a
typical topography of cold-induced shivering is seen. In the early
period of cooling, the shivering activity is mainly confined to the
proximal parts of the extremities, a pattern that is also seen during
emotion. After prolonged cooling the more central parts of the body
such as the thigh will start to shiver, whereas muscular activity
decreases in the distal parts. Thus an "affective" type of shivering can
be differentiated from a "thermal" type.

11.1.2 NON-SHIVERING THERMOGENESIS

This mechanism is an effective means of heat production in the
neonates of various mammalian species, including the human infant
(Brück, 1970a,b, 1971, 1978a,b; Hensel et al., 1973), and in the adult
stage of some mammals after prolonged cold exposure. In adult
humans, non-shivering thermogenesis plays no major role, although it
may be present to some extent (Itoh et al., 1970; Blatteis and
Lutherer, 1976; Doi et al., 1979). Non-shivering thermogenesis is
mediated by the sympathetic nervous system, as can be demonstrated
by a depressive effect of adrenergic β-receptor blocking agents on this
type of heat production, as well as by an increase in non-shivering
thermogenesis after sympathetic stimulation or administration of
noradrenaline (Brück, 1970a; Brück and Wünnenberg, 1970; Brück,
1978a,b; Banet et al., 1978a). The pathway in the brain stem and in the
spinal cord that activates non-shivering thermogenesis has not been
more precisely determined yet.

Non-shivering thermogenesis is a function of the so-called "brown
adipose tissue", a specialized fat tissue that differs from the white fat
by its high metabolic rate, its cells characterized by a more centrally
located round nucleus and numerous small fat droplets ("multilocular
fat"), its richness in mitochondria (Hahn and Novak, 1975), and the
synaptic contact between sympathetic nerve terminals and the cell
membranes (Napolitano, 1965; Bargmann et al., 1968). In rodents
most of the brown adipose tissue is located around the neck and on the

back between the scapulae. In the human neonate, there is no compact interscapular fat pad, but the adipose tissue is more diffusely distributed (Fig. 16.5). The total amount of this tissue makes up as much as 5 to 7% of the body mass in newborn rabbits and guinea pigs and about 1.5% in full-term newborn infants (for references see Brück, 1978b).

In the guinea pig, shivering and non-shivering thermogenesis form a "meshed" control system. The analysis of these two effector mechanisms by Brück and Wünnenberg (1970) has shown that, during external cooling of newborn and cold-adapted guinea pigs, non-shivering thermogenesis is initiated first and shivering occurs in addition only after more severe cooling. However, shivering is immediately evoked if non-shivering thermogenesis is blocked prior to cold exposure by an adrenergic β-receptor blocking agent.

Non-shivering thermogenesis is a function of both the temperature of the body surface and the anterior hypothalamus. Thus the activation of non-shivering thermogenesis by external cooling can be totally suppressed by local heating of the anterior hypothalamic region. In contrast, shivering in the guinea pig is almost independent of the hypothalamic temperature; instead, apart from the surface temperature, it is controlled by the temperature of a circumscribed area in the cervical spinal cord (C6 to T1). This temperature is highly dependent on the heat supplied from the interscapular and cervical brown adipose tissue, which is the site of non-shivering heat production. Thus shivering remains suppressed as long as the capacity of non-shivering thermogenesis is sufficient to maintain the temperature of the cervical spinal cord above a certain level. Further analysis of this control system showed that both shivering and non-shivering thermogenesis appear to be determined in each case by the product of two temperature deviations in the receptor region from the respective reference temperatures (Brück and Wünnenberg, 1970). In the rat non-shivering thermogenesis can be elicited by cooling both the hypothalamus and the spinal cord (Banet et al., 1978a). For further analysis of the control of non-shivering thermogenesis see Chapter 16.

11.1.3 VASOMOTOR CONTROL

The thermoregulatory control of peripheral blood flow in man varies regionally. At least three functionally different regions can be distinguished: (i) extremities (hand, foot, ears, lips, nose), (ii) trunk and proximal limbs, (iii) head and brow (Golenhofen, 1971). Blood flow through the extremities is controlled exclusively via noradrenergic sympathetic fibres. An increase in sympathetic tone causes vaso-constriction and a decrease in tone vasodilatation. In contrast to the extremities, heat-induced reflex vasodilatation in the trunk and proximal limbs is due to an active vasodilatation. This mechanism is cholinergic and acts via the release of bradykinin from the sweat glands. Accordingly, cutaneous regions in which this active heat dilatation is present are the same in which pronounced thermal sweating occurs. Vasomotor nerves exhibit only a slight effect on the

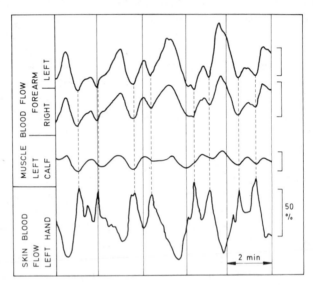

FIG. 11.1. Simultaneous records of skin and muscle blood flow at various sites in a human subject at thermoneutral conditions. (From Golenhofen, 1962.)

forehead, and there is practically no vasoconstriction in response to cold stress. However, vasodilatation does occur together with sweat secretion in response to heat stress (Melchior and Hildebrandt, 1967; Golenhofen, 1971).

In the thermoneutral range there is a continuous oscillation between cutaneous vasconstriction and vasodilatation in human subjects. The period of this oscillation is in the order of 1 min, and the waves are synchronized over the whole body, skin and muscle blood flow varying in antagonistic directions (Fig. 11.1).

During heating of the body in human subjects, the skin blood flow in a limb outside the heated area increases, while at the same time the muscle blood flow can decrease (Barcroft et al., 1955) (Fig. 11.2). This must be taken into account when thermoregulatory changes in cutaneous blood flow are assessed by plethysmography of muscular parts of limbs (e.g. forearm) or by measuring total blood flow of extremities (Wyss *et al.*, 1974, 1975; Wenger *et al.*, 1975b; Proppe *et al.*, 1976; Johnson *et al.*, 1976; Nadel *et al.*, 1979).

A study by Melchior and Hildebrandt (1967) of regional cutaneous blood flow in man has revealed different topographic patterns when

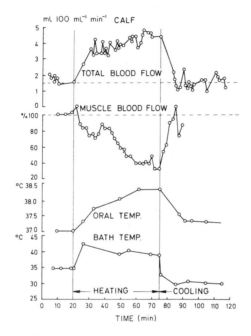

FIG. 11.2. Total blood flow in the human calf and blood flow in the gastrocnemius muscle when the remaining part of the body was heated and cooled in a bath. (From Barcroft *et al.*, 1955.)

the body was passively heated or when heat was produced by active work. As is known, passive heating causes particularly high increases in blood flow in the extremities, such as the hand and foot. However, when subjects were exercising with 90 W for 20 min, and heat load increased, cutaneous vasodilatation was confined to the profusely sweating areas of head, trunk and proximal limbs, while the blood flow of the extremities remained almost unchanged (Fig. 11.3). The average increase in skin blood flow was 250% in the forehead and the upper arm, 160% in the chest, but only 12% in the thumb. In this situation the vasomotor tone of the extremities is obviously involved in regulation of blood pressure rather than in thermoregulation. At the end of the exercise, the blood flow in forehead and the upper arm fell immediately, whereas the blood flow in the thumb showed a considerable increase. The vasomotor pattern was then similar to that observed in passive hyperthermia.

A direct comparison of sympathetic activity and cutaneous blood flow during thermal stimulation of the CNS was made by Simon

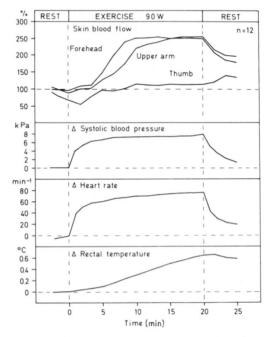

FIG. 11.3. Skin blood flow at various sites, systolic blood pressure, heart rate and rectal temperature in a human subject during rest and exercise. (From Melchior and Hildebrandt, 1967.)

(1971). When either the hypothalamus or the spinal cord of anaesthetized rabbits was heated, a vasodilatation in the ear and a decrease in mass discharge of the sympathetic nerve branch supplying the ear were seen. Cooling either area of the CNS had the opposite effect. Furthermore, a regional differentiation of vasomotor activity occurred, in that the splanchnic nerves showed the opposite response as did the branches to the ear, i.e. an increase in activity during central warming and a decrease during cooling. This reciprocal regional vasomotor cutaneous and splanchnic outflow to blood vessels can be considered as a mechanism providing circulatory homeostasis under the conditions of thermally induced changes of skin blood flow.

Under thermoneutral conditions, spontaneous fluctuations both in efferent sympathetic activity and ear blood flow in the rabbit were seen (Iriki and Hales, 1976), even when skin and rectal temperature remained constant (Fig. 11.4). This spontaneous fluctuation had a longer period of oscillation (about 3 min) than that observed in humans (Fig. 11.1) but the difference may be due to anaesthesia.

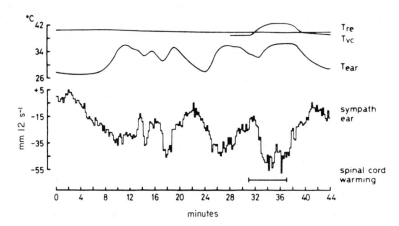

FIG. 11.4. Integrated cutaneous efferent sympathetic activity (sympath. ear), and temperatures of the ear (T_{ear}), vertebral canal (T_{vc}) and rectum (T_{re}) of a rabbit lying on a pad at 37.8 to 38.1 °C with ambient dry bulb temperature 23.8 to 25.1 °C. (From Iriki and Hales, 1976.)

The findings in rabbits were confirmed in cats (Ninomiya and Fujita, 1976). There was an antagonism between the afferent activity in the sympathetic nerves supplying the skin and those supplying the kidney. The level of cutaneous sympathetic activity was a function of both skin temperature and hypothalamic temperature, but, in contrast to other findings on cutaneous blood flow, (cf. p. 119) the effect of hypothalamic temperature was opposed to that of skin temperature.

Cutaneous veins possess a specialized thermoregulatory function: in the dog, for example, body cooling causes a generalized cutaneous venoconstriction and body heating a venodilatation, which cannot be observed in the splanchnic capacity vessels (Webb-Peploe and

Shepherd, 1968a,b,c; Webb-Peploe, 1969a,b). These venomotor reactions, mediated by adrenergic nerves, are part of the general thermoregulatory control of the cutaneous circulation that operates to restrict or increase heat loss from the skin according to thermal requirements of the organism; the cutaneous venoconstriction with body cooling is thought to decrease heat loss by reducing the venous surface area and directing venous flow through the venae comitantes, whose transfer of heat takes place with the accompanying artery (Vanhoutte and Shepherd, 1971). Such a countercurrent system carries some of the arterial heat back into the body. Since the deep veins in the limbs do not participate in the venomotor reflexes (Zelis and Mason, 1969), this facilitates the shift of blood from the superficial to the deep veins. The effectiveness of these venous thermoregulatory mechanisms is greatly enhanced by local thermal influences modulating the reactions of the veins to the changes in central venomotor outflow (p. 165).

11.1.4 SUDOMOTOR CONTROL

When the whole body is exposed to ambient heat, sweating appears on the lower extremities with recruitment proceeding in a rostral direction. Heating only the upper cutaneous areas markedly decreased the sweat-recruitment time or reversed the normal order of recruitment, whereas heating only the lower half elicited the usual recruitment pattern (Wurster et al., 1969). A particularly high sweating rate is found in the forehead (Houdas et al., 1972). McCook and his colleagues (McCook et al., 1970) assume that afferent pathways from individual areas of the skin exist that influence the sweat glands of that area and possibly adjacent areas. These pathways may pass up the cord with mediation within the brain or may activate reflexes limited to the cord and sympathetic trunk.

Sweat glands in various mammals, such as Canidae, Bovidae, Equidae, and Prosimii, are controlled by adrenergic mechanisms, whereas eccrine thermal sweating in man is controlled via cholinergic sympathetic pathways (Robertshaw, 1971; Robertshaw et al., 1973), although adrenergic sweating can be found in humans as well (McCook et al., 1970; Allen and Roddie, 1972). In the human forearm neural transmission to the sweat glands can be blocked by small doses of atropine (Foster and Weiner, 1970; Gibinsky et al., 1973). Of anti-adrenergic substances examined, only high doses of guanethidine and phentolamine produced any consistent and significant measure of blockade. It can be concluded that there are only cholinergic fibres innervating the sweat glands, and any inhibitory action by anti-adrenergic substances is due to interaction with cholinergic receptors in the glands. The sweating threshold (5% increase) for intradermally administered acetylcholine was $10^{-3}\,\mu$g given in 0.1 ml. When the skin temperature was 36 °C, the maximal acetylcholine response was then the same as the maximal thermal response (Foster, 1971).

11.2 Local Effects of Temperature

Local skin temperature influences the peripheral effectors of thermo-regulation, such as cutaneous blood flow and sweat secretion. These local mechanisms modify the effect of efferent sympathetic impulses on the blood vessels and sweat glands. Thus the skin temperature has a twofold influence: (i) via the cutaneous thermoreceptors, afferent nerves, and integrating centres; (ii) via the direct effect of temperature on the peripheral control actions.

11.2.1 CUTANEOUS BLOOD VESSELS

Isolated and denervated limb areas continue to respond to thermal stimuli by vasomotor adjustments. Direct effects of temperature on the blood vessels, the formation of vasoactive substances in the skin, such as bradykinin (Gautherie, 1971), and local reflexes have been considered as explanations for this phenomenon (Keatinge, 1970).

Keatinge (1970), working with isolated segments of the a. ulnaris of the ox, showed that the constrictive effect of catecholamines is abolished at temperatures below 10 °C. In a temperature range between 41 and 30 °C, the tension caused by noradrenaline (NA) increased up to fivefold with falling temperature. As the temperature fell still further, the development of tension diminished and at 15 °C it again reached the initial level (Sams and Winkelmann, 1969). According to the experiments on isolated arteries, maximal sensitivity to sympathetic constrictive stimuli would be expected at a local temperature of 30 °C. However, in the intact dog the maximal constrictive effect of sympathetic impulses on cutaneous veins was not reached at temperatures as low as 17 °C (Webb-Peploe and Shepherd, 1968c). In any case, if the local temperature sinks to very low levels, the central constrictive stimuli must gradually become ineffective again. These results can explain the observations that, under the influence of local heating of the forearm, a considerable increase in cutaneous blood flow is seen that also occurs in patients with cervical sympathectomy or complete paralysis of the brachial plexus.

The direct effect of temperature on blood vessels has also been advanced as an explanation for the so-called cold vasodilatation (Lewis' hunting reaction). This phenomenon appears when, for example, a hand is immersed in ice water. At first, maximal vasoconstriction sets in and the temperature of the finger tips falls to a level close to that of the water. After some time, a sudden intense increase in the finger temperature takes place due to vasodilatation of cutaneous vessels. If the hand remains in cold water, this series of events is periodically repeated (Fig. 11.5). The reaction described here takes place in a similar manner after sympathectomy or complete

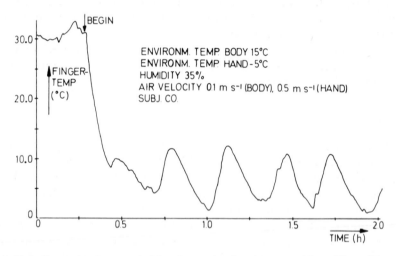

FIG. 11.5. Example of a recorded Lewis reaction in a human subject. (From Werner, 1977.)

denervation (for references see Keatinge, 1970; Werner, 1977b). However, the Lewis reaction is also dependent on the general thermoregulatory state of the body. For example, the amplitude of the oscillations increases with increasing environmental temperature of the body, and after a cold load of the body the Lewis reaction is suppressed in some individuals (Werner, 1977a). The large undamped oscillations of the Lewis reaction remain to be explained, since, from a theoretical point of view, the blood flow might also reach a steady state defined by an equilibrium between cold paralysis of blood vessels and vasoconstrictive impulses.

The tone of cutaneous veins is governed both by central thermoregulatory and by local thermosensitive mechanisms. According to Webb-Peploe and Shepherd (1968a,b,c), local cooling of cutaneous veins in the dog increased their constrictor response to adrenergic sympathetic stimulation, while local warming had a depressive effect. When a vein was locally perfused at 42 and 17 °C, the responses to sympathetic stimulation were 40% and 200%, respectively, compared with the responses at 37 °C. Similar results were obtained for the constriction caused by NA (Webb-Peploe and Shepherd, 1968c). The venomotor tone showed a reciprocal linear relationship to the central body temperature over the range of 34 to 40 °C, and this central effect interfered with the response to local temperture. For example, a decrease in central temperature from 38 to 34 °C increased the venoconstrictor response to local cooling by a factor of 10 or more (Webb-Peploe and Shepherd, 1968b).

Further analysis of the local temperature sensitivity of the venomotor reaction has shown that this effect persisted during electrical stimulation of isolated and denervated veins; without stimulation, isolated veins did not respond to temperature changes (Vanhoutte and Leusen, 1969; Vanhoutte and Shepherd, 1970). Potentiation seen with cooling was still present after inhibition of both neuronal reuptake and enzymatic breakdown of NA (Vanhoutte and Shepherd, 1969; Webb-Peploe, 1969a,b), after α- and β-adrenergic inactivation, and during venoconstriction caused by substances other than NA (Vanhoutte and Shepherd, 1970). Thus neither the sympathetic nerves nor functional adrenergic effectors are essential for the observed potentiation with cooling. Temperature may be acting in a non-specific way, without interference from the classical receptor mechanisms, directly on the smooth muscle cells (Vanhoutte and Shepherd, 1971).

In human subjects, the interference of central and local effects of temperature on the venomotor tone was confirmed. Cutaneous venoconstriction caused by a deep breath, by mental arithmetic, or by exercise is released by local warming of the skin. The same holds true for venomotor responses induced by changes in core temperature or in average skin temperature; these responses are as well influenced by the local temperature of the veins (Shepherd and Webb-Peploe, 1970; Rowell et al., 1971a,b; Zitnik, 1971).

11.2.2 SWEAT GLANDS

Bullard and his colleagues (Bullard et al., 1970; Elizondo and Bullard, 1971) have demonstrated in humans that the rate of eccrine sweat secretion in a small skin area is appreciably changed with local temperature variations. This effect is strictly limited to the area affected by the temperature change. Local heating was effective only in the presence of an adequate central drive. The effect of local temperature changes on the eccrine sweat glands was reduced after blocking neuroglandular transmission by hemicholinium-3; similar effects were produced by arterial occlusion. An increase in the response to local heating was seen after treatment with physostigmine, leading to an increased release of the neuroglandular transmitter acetylcholine. Substances that act directly on the sweat glands, such as methylcholine chloride and pilocarpine, led to an increased sweat secretion but this effect was not significantly enhanced by local heating (MacIntyre et al., 1968; Bullard et al., 1970). The conclusion drawn from these experiments was that the site affected by temperature is the neuroglandular junction. Although arterial occlusion and local heating were found to alter the sweating rate, they did not significantly change the relationship between sweat production and electrolyte excretion (Elizondo et al., 1972). Local heating of the skin might also activate mechanisms other than neuroglandular transmission, since increases in sweat secretion may occur even in denervated and atropinized areas; the nature of these events has not yet been established (Ogawa, 1970).

The effect of local skin temperature (T_{sl}) on sweat secretion is of considerable quantitative significance. One can describe the local sweating rate (E_l) by the equation

$$E_l = E_c \cdot L \tag{18}$$

where E_c is the central drive dependent on core and average skin temperature and L is a factor expressing the effect of local temperature. The relationship between L and local skin temperature (R_{sl}) is an exponential function, with T_{sl} included in the exponent (Nadel et al., 1971; Nadel and Stolwijk, 1971; Stolwijk et al., 1971). Thus the local sweating rate follows the equation

$$E_l = E_c e^{(T_{sl} - 34)/10} \tag{19}$$

As a result of the local factor, sweat secretion practically ceases at low skin temperatures, even if the central drive is maximal. Another observation that might be explained by local effects of temperature is the high sweat rate in skin areas overlying active muscles (Wells and Buskirk, 1971).

The whole body sweating response (m_{sw}) can be described by the equation (Nadel et al., 1971; Nadel and Stolwijk, 1973)

$$m_{sw} = \Phi\psi[\alpha'(T_{es} - 36.7) + \beta' (\bar{T}_s - 34.0)] e^{(\bar{T}_s - 34.0)/10} \tag{20}$$

Φ is a factor describing the proportion between sweating rate and sweating drive, ψ is a function of the state of acclimation, T_{oe} is the oesophageal and T_s the mean skin temperature. α' and β' are proportional control constants.

The factor Φ depends on skin wettedness (Henane, 1972; Nadel and Stolwijk, 1973). When the skin was frequently wiped dry at otherwise equal thermal states of the body, the sweating rate increased considerably. This means that the factor ψ increases with decreasing skin wettedness. It has been suggested that possibly an increased osmotic pressure on the skin surface might enhance the activity of sweat glands (Nadel and Stolwijk, 1973).

12
Thermal Comfort in Man

12.1 Thermal Comfort and Temperature Sensation

12.1.1 PHENOMENOLOGY AND SEMANTICS

Under the conditions of daily life we do not clearly distinguish between thermal comfort and temperature sensation. A closer investigation, however, reveals that both kinds of experience can be separated phenomenologically and physiologically. Temperature sensation is a rational experience that can be described as being directed towards an objective world, as expressed by the statement: "It is cold." Thermal comfort is an emotional or affective experience referring to the subjective state of the observer as expressed by the statement: "I feel cold." The verbal scales describing temperature sensation are based on the terms "cold" and "warm", whereas thermal comfort and discomfort can be characterized by the terms "pleasant" and "unpleasant" (Cabanac, 1969, 1972, 1979). According to the Oxford Dictionary, comfort means "satisfaction", "enjoyment" and "pleasure". The Concise Oxford Dictionary defines "pleasure" as "agreeable", and "agreeable" as "pleasing".

Neither do investigators of thermal comfort always clearly discern between thermal comfort and temperature sensation. Even in more recent publications, thermal comfort is sometimes referred to as "thermal sensation" (e.g. McIntyre, 1976; Ballantine et al., 1977). This confusion is also reflected by various comfort scales (for historical references see Gagge, 1979). For example, the seven-point scale used by the American Society for Heating, Refrigeration and Air-Conditioning Engineers (ASHRAE, 1966) contains expressions of cold and warmth without mentioning whether these are pleasant or unpleasant (Table 12.1). However, according to ASHRAE's definition, thermal comfort is "that condition of mind which expresses satisfaction with the thermal environment". A certain attempt to

168

distinguish between thermal comfort and temperature sensation can be seen (Table 12.1) in the seven-point Bedford scale (Bedford, 1936). Here the expressions "too" and "much too" indicate that the observer is not satisfied with his thermal environment.

TABLE 12.I

Comfort expressions

Bedford (1936)	ASHRAE (1966)
1. Much too cool	1. Cold
2. Too cool	2. Cool
3. Comfortably cool	3. Slightly cool
4. Comfortable and neither cool nor warm	4. Neutral
5. Comfortably warm	5. Slightly warm
6. Too warm	6. Warm
7. Much too warm	7. Hot

Among physiologists, Ebbecke (1917, 1948) was first to distinguish between temperature sensations and thermal affects; thermal sensations were assumed to be correlated with the activity of cutaneous thermoreceptors, whereas thermal affects were assumed to depend on thermoregulatory reflexes ("Reflexempfindungen") originating from the heat regulation centre ("Wärmezentrum"), a view that has largely been confirmed half a century later. A clear classification of both types of experiences was given by Winslow and Herrington (1949), Gagge et al. (1967), and Hardy (1970), who used separate scales for comfort and sensation, respectively (Table 12.II).

TABLE 12.II

Expressions of comfort and thermal sensation

Scale of discomfort	Scale of thermal sensation
1. Comfortable	1. Cold
2. Slightly uncomfortable	2. Cool
3. Uncomfortable	3. Slightly cool
4. Very uncomfortable	4. Neutral
	5. Slightly warm
	6. Warm
	7. Hot

From Hardy (1970).

12.1.2 DIFFERENTIATION BETWEEN COMFORT AND SENSATION

Various experimental procedures have revealed marked dissociations between thermal sensation and comfort. Chatonnet *et al.* (1964) exposed human subjects to a sudden current of cold air at 4 or 18 °C

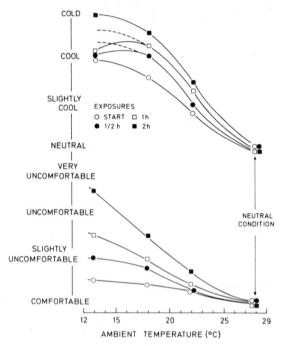

FIG. 12.1. Discomfort and thermal sensation in different temperatures after ½, 1 and 2 hours of exposure. (From Hardy *et al.*, 1971.)

for 30 to 45 min, and the subjects reported on their thermal sensations. At first there was a distinct but not unpleasant sensation of cold, but gradually the cold sensation intensified and became very unpleasant. These unpleasant sensations were accompanied by intermittent bursts of shivering, after which the level of unpleasantness was reduced.

The results of similar experiments (Gagge *et al.*, 1967; Hardy *et al.*, 1971; Stolwijk, 1979) are shown in Fig. 12.1. After a sudden exposure to low ambient temperature, cold sensations start very soon, thus indicating a dominant role of skin temperature for this perception. Reports of discomfort occur much later, indicating that a change in body heat content or internal body temperature is an important prerequisite for comfort sensation. Body temperatures near normal can produce reports of discomfort, if the body temperature is changing in a direction away from the normal. Conversely, if the body temperature is in an abnormally high or low range, any exposure which tends to return the abnormal body temperature towards normal will be perceived as comfortable, and any exposure which will

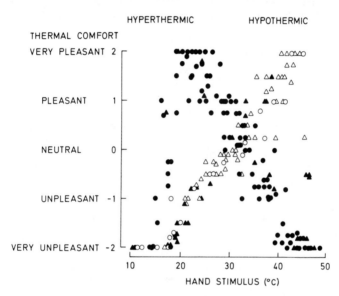

FIG. 12.2. Affective responses given by a human subject to 30 s thermal stimulations of the left hand. Each point corresponds to a stimulation. Hypothermia: △ bath 33 °C, ○ bath 38 °C, deep body temperatures 36.3 to 36.6 °C. Hyperthermia: ▲ bath 38 °C, ● bath 30 °C, deep body temperatures 37.0 to 37.8 °C. (From Cabanac, 1969.)

cause the same body temperature to move even further away from normal will be perceived as very uncomfortable. There was a marked difference between men and women in their discomfort reports: women reported earlier and more severe discomfort in the cold, and men reported earlier discomfort in the heat (cf. Chapter 16).

Chatonnet and Cabanac (1965) and Cabanac (1969) systematically changed the core temperature of human subjects by immersing them in baths of various temperatures. When the subjects were hypothermic, warm stimuli applied to the hand were experienced as pleasant, cold stimuli as unpleasant (Fig. 12.2). The opposite responses were observed in hyperthermic subjects. By judging the intensity of comfort or discomfort in a phenomenal scale, it was possible to establish the subject's core temperature with a remarkably high degree of accuracy. In contrast to the drastically changing affective responses, the cold stimuli, in terms of temperature sensation, were always experienced as "cold", and the warm stimuli as "warm".

These experiences indicate that thermal comfort is influenced both by core and skin temperature, while temperature sensations seem to depend mainly on skin temperature. Benzinger (1970, 1979) has observed that the threshold of cold sensation was 34.5 °C average skin

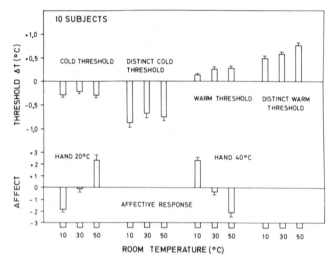

FIG. 12.3. Cold and warm thresholds of the hand and affective responses to thermal stimulation of the hand with 20 and 40 °C at room temperatures of 10, 30 and 50 °C. Average values of 10 subjects; bars indicate standard error of mean. (From Hensel, 1977.)

temperature, and virtually independent of tympanic temperature between 36.0 to 37.7 °C.

In order to further investigate this question, local thresholds of warm and cold sensation as well as thermal comfort elicited from the same area of stimulation were tested in subjects exposed for 1 h at room temperatures of 10, 30 and 50 °C, respectively (Hensel, 1974c, 1977, 1979a,b). At 10 °C ambient temperature the average skin temperature was 27.3 °C and the rectal temperature 36.5 °C, while at 50 °C ambient temperature the respective values were 36.8 °C and 37.5 °C. The arm was thermally insulated from the ambient conditions, the hand being placed on a large thermode. Linear thermal stimuli of $+1.5\,^{\circ}\mathrm{C\,min}^{-1}$ and $-1.5\,^{\circ}\mathrm{C\,min}^{-1}$ were applied. In addition, the other hand was immersed alternatively in stirred water baths of 20 and 40 °C, thereby judging the degree of pleasantness on a subjective scale.

Figure 12.3 shows that local cold thresholds on the hand are not considerably altered by ambients of 10 and 50 °C, whereas the

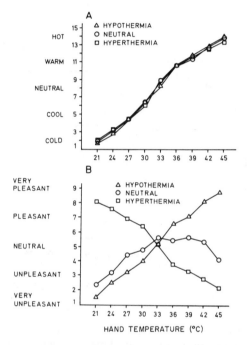

FIG. 12.4. A: Group mean category judgments of thermal intensity as a function of stimulus temperature of the hand under three conditions of internal body temperature. B: Group mean category judgments of thermal pleasantness as a function of stimulus temperature under three conditions of internal body temperature. Each point represents the mean of four judgments by each of four subjects. (From Mower, 1976.)

pleasantness of cold stimuli of 20 and 40 °C is drastically modified by various room temperatures. In contrast to the constant cold thresholds, the warm thresholds increase to a certain extent with increasing room temperature. This may be interpreted as a kind of "masking" of the warm thresholds by the thermal afferents from the whole body. These findings indicate that the sensory channels for cold and for warmth may have somewhat different properties.

Mower (1976) tested the perceived magnitude of 9 temperature stimuli ranging from 21 to 45 °C applied to the hand of subjects that were made hypothermic or hyperthermic, respectively, by immersion in water baths. As shown in Fig. 12.4, there was practically no change in perceived magnitude with the general thermal state of the observer. In contrast, thermal pleasantness was highly dependent on body temperature.

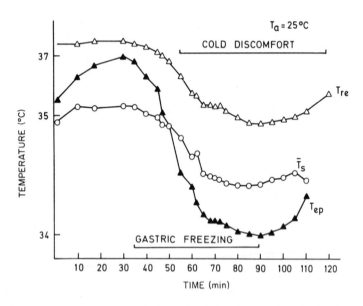

Fig. 12.5. Body temperatures and thermal comfort in a patient during intragastric cooling to −4 °C by means of a balloon. (From Chatonnet et al., 1966.)

The present results indicate that whereas thermal pleasantness is markedly affected by deviations of internal body temperature from normal (Fig. 12.5), the intensity of temperature sensations is independent of such manipulations. Perceived thermal intensity depends on signals from the peripheral thermoreceptors and is independent of stimulation of the thermal interoceptors. In contrast, thermal pleasantness is the result of an interaction of signals from both central and peripheral receptors.

The pleasantness of dynamic warm stimuli applied to the forehead is modified by the mean skin temperature, even when the internal body temperature remains unchanged (Marks and Gonzalez, 1974). This holds also for the pleasantness of static thermal stimuli applied to the hand (Beste, 1977; Beste and Hensel, 1978). When the palm is

kept at 25 °C and the room temperature slowly changed so as to shift
the mean skin temperature from 31 to 37 °C, the thermal affect elicited
from the hand changes from unpleasant to pleasant. Since the rectal
temperature remains practically constant, this affective change must
be ascribed to cutaneous thermal afferents alone. With constant hand

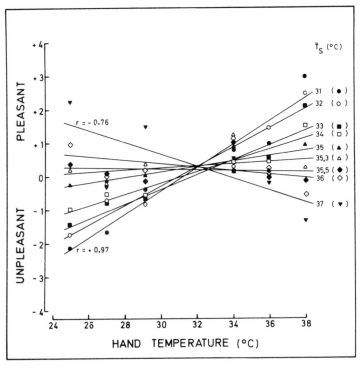

FIG. 12.6. Pleasantness of various constant temperatures of the hand at various
average skin temperatures. (From Beste, 1977.)

temperatures of 38 °C, the thermal affect changes from pleasant to
unpleasant during a shift in average skin temperature from 31 to
37 °C. The diagram in Fig. 12.6 shows the pleasantness of constant
temperatures of the hand between 25 and 38 °C at average skin
temperatures between 31 and 37 °C. It is remarkable that in the
neutral range of average skin temperature, any constant temperature
of the hand between 25 and 38 °C is rated neither pleasant nor
unpleasant (neutral = 0), whereas distinct pleasantness or unpleasantness is observed only when the average skin temperature
deviates considerably from the thermally neutral range. These findings agree in part with observations by Mower (1976).

12.2 Physiological Conditions of Thermal Comfort

In the steady state the physiological conditions of general thermal comfort or thermal neutrality can be described as follows: (1) internal body temperature 36.6 to 37.1 °C, i.e. generally on the warm side of the set point (36.6 °C) for the internal body temperature; (2) mean skin temperature 33 to 34.5 °C for men and 32.5 to 35 °C for women; (3) local skin temperature is variable over the body but generally between 32 and 35.5 °C; (4) temperature regulation active and

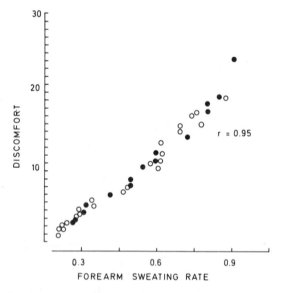

Fig. 12.7. Relation between magnitude estimates of discomfort and forearm sweating rate in mg min^{-1} cm^{-2} on days one (open circles) and six of exposure to high humidity. (Modified from Gonzalez et al., 1973.)

completely accomplished by vasomotor control of blood flow to the skin—no sweating or shivering present—with blood flow to hands and feet high, producing reversed thermal gradients along the extremities (Hardy, 1970).

It is well established now that thermal comfort reflects an integration of various thermal drives (Cabanac, 1969, 1972, 1979; Gagge et al., 1969; Hardy et al., 1971; Hensel, 1973c, 1977, 1979a,b). According to Hardy (1970), the physiological correlate of thermal comfort is a composite of (a) the signals evoking temperature sensations; (b) the signals for temperature regulation; and (c) those sensations arising from the thermoregulatory activities of vasomotor control of skin blood flow, sweating, and shivering.

A close relationship between thermal comfort and temperature regulation can be derived from investigations in which autonomic responses and comfort were assessed simultaneously. Thus the magnitude estimate of warm discomfort was found to be closely correlated with the forearm sweating rate (Gonzalez *et al.*, 1973), as shown in Fig. 12.7. Since the sweating rate from a local skin area is a good reflection of the central sweating drive, which itself is a function of weighted mean body temperature (p. 128), it is apparent that warm discomfort is related to the body temperature in a similar manner as is the physiological response.

However, there are findings which are not compatible with the view that the drives for thermal comfort and temperature regulation are identical. When subjects immersed in a water bath could regulate the water temperature according to their thermal comfort, the resulting deep body and skin temperatures during physical exercise caused an increase in sweating rate and heat conductance of the body (Bleichert *et al.*, 1973). Thus, during exercise, thermoregulatory responses increase as a function of oxygen consumption, although the subjects are at comfort.

According to Bleichert *et al.* (1973) the different functions of the systems regulating sweating rate and thermoregulatory behaviour may arise from different weighting factors of oesophageal temperature (T_{oe}) and average skin temperature ($\bar{T}_s$) as input variables in both systems. In the equation for the control action (y)

$$y = a_0 + a_1 T_{oe} + a_2 \bar{T}_s \tag{21}$$

the ratio $a_1 : a_2$ for $y = 0$ is about 12 to 15 in the case of sweating rate and about 4 in the case of behaviour. The different weighting factors can be derived from Fig. 12.8. The diagram correlates T_{oe} and $\bar{T}_s$ for the sweating threshold and for the threshold of warm discomfort, respectively. As can be seen, both functions have different slopes which means that their set point conditions are different. At high $\bar{T}_s$ and low T_{oe}, the set point for sweating and discomfort is virtually identical, whereas at $T_s = 27\,°C$ the sweating threshold corresponds to $T_{oe} = 36.75\,°C$ and the threshold for warm discomfort to $37.8\,°C$.

From these results it cannot be concluded with certainty whether a non-thermal factor due to exercise is involved in the difference in set points for autonomic and behavioural responses. As to autonomic thermoregulation, it is unlikely that the set point is shifted by physical

work (p. 205). Neither is there an indication for a set-point displacement for thermal comfort, since the data points in Fig. 12.8 during the
working period (active heating) are nearly the same as those for the
subsequent resting period (passive heating).

The possible neural processes involved in temperature sensation,
thermal comfort and temperature regulation are summarized in the
schematic diagrams in Fig. 12.9. We can assume that the signals for

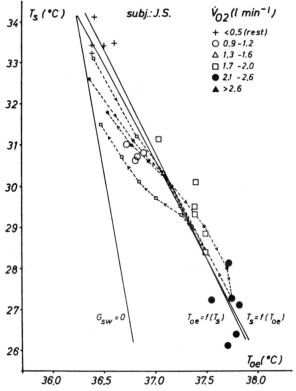

FIG. 12.8. Relationship of skin temperature (T_s) to oesophageal temperature (T_{oe}) at
thermal comfort during rest and exercise. For further explanation see text. (From
Bleichert *et al.*, 1973.)

comfort and discomfort originate from thermosensitive structures in
the central nervous system, other deep body thermosensors,
cutaneous thermoreceptors, and perhaps structures involved in
shivering, sweating and vasomotion. The integrated signals from
various internal and external thermosensors give rise to general
thermal comfort but are also involved in local comfort. The same
holds for the processes of thermoregulation. The diagram also

accounts for the possibility of a dissociation between autonomic thermoregulation and thermal comfort. As to the thermoreceptors in the skin, they contribute on the one hand to the integrated signals for thermoregulation and thermal comfort, but on the other hand their signals are mediated through separate pathways involved in temperature sensation. The local cold sensations are practically independent of the subject's general thermal state, whereas a slight influence may be assumed for the warm sensation.

The biological importance of thermal comfort can be seen in the fact that it motivates behavioural responses according to the load

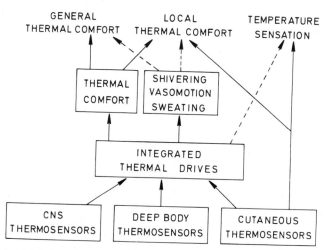

FIG. 12.9. Physiological conditions for temperature sensation, thermal comfort and temperature regulation. For further explanation see text.

error of thermoregulation. On the other hand, the temperature sensations that are mediated through separate pathways, and therefore are largely independent of the subject's thermoregulatory situation, allow one to judge the thermal environment more objectively and thus to predict what might happen to the organism if it were exposed to that particular environment.

If one agrees that comfort and discomfort depend on the load error $(T - T_0)$, the measurement of thermal comfort in response to given stimuli will provide a possibility to establish the set temperature. This approach has been applied to determine the set point during the diurnal and the menstrual cycle, during physical work, and in the course of fever (Chapter 14).

It should be mentioned that the temperatures of the body are not the only factors that influence the experience of thermal discomfort. For example, it has been shown that there was a good relation between the magnitude estimate of discomfort and the skin wettedness (w), which is the fraction of the total body area covered by sweat and is mathematically equivalent to the evaporative rate divided by the maximal evaporative power of the environment (E_{max}). In transient conditions when T_{oe}, $\bar{T}_s$ and sweating rate are not extremely high, warm discomfort may be great if E_{max} is relatively low and w is high.

Thermal discomfort may also be influenced by a number of other non-thermal factors, such as central circulatory dynamics and heavy exercise. However, the relative importance of these factors and the circumstances under which they come into action are not well understood as yet (Nadel, 1979).

12.3 Local and General Thermal Comfort

It has been said that thermal comfort reflects an integration of thermal afferents from various sites of the body. However, this integration is not unlimited. If the local temperature of a body area deviates to a certain extent from the mean temperature, local comfort or discomfort can be experienced in this area, independent of general thermal comfort. For example, cold feet can be discomfortable, in spite of the fact that the remaining part of the body is comfortable, or even uncomfortably warm. Likewise, large temperature differences between the upper and lower part of the body, or between the frontal and dorsal part, may be uncomfortable, even if the average skin temperature would correspond to a value that otherwise would be comfortable.

Discomfort aroused by thermal exposure of the whole front of the body grows as a power function of the increase or decrease in operative temperature from the temperature that feels comfortable (Stevens et al., 1970). For discomfort of warmth, the exponent was found to be $n = 0.7$, that is less than half as large as the exponent of $n = 1.7$ for the discomfort in cold.

Local thermal discomfort can be created by asymmetric thermal radiation (Olesen et al., 1973), draught (Ostergaard et al., 1974; Fanger, 1977, 1979; McIntyre, 1979), cold or warm floor (Olesen, 1977), non-uniform clothing, and by vertical air temperature difference. For sedentary subjects wearing light clothing, the limit for thermal comfort was established when vertical air temperature gradients between head and ankle were applied (Olesen *et al.*, 1979). The local discomfort caused by vertical air temperature differences may either be warm discomfort at the head or cold discomfort at the feet (or both). It turned out that the percentage of subjects dissatisfied with the comfort conditions was a function of the air temperature gradient (Fig. 12.10). For 180 min occupancy, the air temperature difference between head and ankle should be less than 2.8 °C if it is

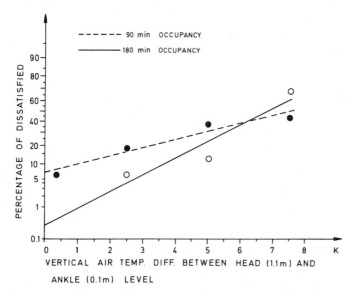

FIG. 12.10. Probit analysis of local thermal discomfort as a function of vertical air temperature differences. (From Olesen *et al.*, 1979.)

accepted that 5% of the subjects feel uncomfortable. From similar experiments (Lebrun and Marret, 1979; Löfstedt, 1979) in sedentary subjects it was concluded that the air temperature around the body has more influence than the air temperature around the feet, and that the local sensations of discomfort at the feet lastly seem to be due to thermoregulatory responses much more than to the effects of local temperature stimuli. This may be seen in connection with Hardy's (1970) hypothesis that vasomotor activity is involved in the experience of thermal comfort.

Recently the hypothesis of a vasomotor component in local thermal comfort was tested by comparative measurements of skin temperature and cutaneous blood flow in the foot of human subjects (Issing and

Hensel, 1979). Room temperature and foot temperature could be varied independently in various combinations. The subjects estimated local temperature sensations as well as local and general thermal comfort. There were considerable interindividual differences in cutaneous blood flow as function of average skin temperature and local foot temperature. Identical foot temperatures were associated with approximately the same degree of local comfort, independent of a high or low rate of skin blood flow. On the other hand, local comfort could change drastically with foot temperature in spite of a constant blood flow. The results do not support the hypothesis that vasomotor influences other than changes in tissue temperature may be involved in local thermal comfort or discomfort.

Local cold discomfort can be elicited from the face by cold wind at speeds between 4 and 7 ms^{-1} (LeBlanc et al., 1976). There is a good correlation between the drop in skin temperature and the subjective evaluation of cold discomfort. Furthermore, cooling the face elicits bradycardia which also corresponds well with cold discomfort. Consequently, the drop in skin temperature, reflex bradycardia, and subjective evaluation are parameters which are directly affected by cold wind and can be used as adequate indicators of the degree of local discomfort.

12.4 Assessment of Thermal Comfort in Practice

In the study of thermal comfort, the engineering profession had originally taken the lead (for historical references see Gagge, 1979). With the development of heating and air conditioning it became necessary to define the conditions under which an artificial climate is comfortable or acceptable to humans at various occupations and with various clothing. The aim is to develop comfort equations that are able to predict the effect of a given combination of air temperature, radiation temperature, humidity, air velocity, and certain local factors, such as floor temperature, spatial temperature gradients, asymmetric temperature fields etc. upon human comfort.

The comfort scales have been refined by probit analysis (Fanger, 1973a; Ballantyne et al., 1977; Fanger and Valbjørn, 1979). By this method the degree of comfort is expressed by the percentage of subjects who accept a given thermal situation as comfortable.

Since this chapter is mainly concerned with physiological aspects, it would be beyond its scope to discuss in detail the comfort conditions expressed in physical parameters of man's thermal environment. A great deal of work has been invested in these studies, which become increasingly important with the growing problem of energy conservation. There are a number of comprehensive books (Fanger, 1970, 1973a), symposia (Houdas and Guieu, 1978; Durand and Raynaud, 1979; Fanger and Valbjørn, 1979) and various reviews (e.g. Fanger, 1973b,c, 1976; Olesen, 1977) which may be consulted for the assessment of man's thermal comfort in practice and for information about human comfort under various conditions, such as age, sex, race, climate, season, physical fitness, and other personal factors. Some of these are discussed in Chapters 15 and 16.

Considering the following six variables: (1) metabolic rate (activity), (2) air temperature, (3) air humidity, (4) mean radiant temperature, (5) air velocity and (6) clothing, a comfort equation has been developed (Fanger, 1970) which allows to predict whether a given combination of these factors is comfortable or not. Furthermore, if a number of factors are given, one can calculate the value of the remaining factors necessary for comfort conditions.

The ambient temperature preferred for thermal comfort, when determined under controlled steady state conditions, appears to be remarkably uniform. Recently there has developed an increased interest in responses to temperature which are either slowly changing around or near the preferred value, or constant, slightly above or below preferred temperatures. One of the practical implications is the question whether a slow drift of temperature from the preferred value can be acceptable in order to save heating energy for certain periods of time.

When sedentary subjects were exposed to slowly changing ambient temperatures of 0.5 to $1.5\,°C\,h^{-1}$, a certain deviation from the preferred temperature was accepted (Berglund, 1979). The accepted deviation was larger when the temperature change was slower (Fig. 12.11). For example, with the slowest cooling of $0.5\,°C\,h^{-1}$ a decrease from 25 to 21 °C ambient temperature was accepted by 100% of the subjects wearing summer clothing. There is a certain dissociation between discomfort and acceptance; although the environment becomes slightly uncomfortable, it is still accepted.

A comparative study of different heating systems (radiator, convector, heated ceiling, floor heating, warm air, skirting board) in a thermally insulated room led to the result that all the systems tested were able to create a comfortable thermal environment according to the subjective voting scales (Olesen and Thorsauge, 1979), provided that the basic requirements for comfort conditions are met.

To achieve true comfort the temperature should be maintained at a level which enables thermal neutrality to be ascertained for the body in general, and no local thermal discomfort should occur. As Fanger (1977) points out, local cold discomfort caused by air velocity, a cold floor, or by asymmetric radiation from a large window is of major importance for energy conservation. Occupants experiencing local cold discomfort will require a higher ambient temperature level. This will decrease the local discomfort but will at the same time shift the general comfort to the warm side. The higher the temperature level,

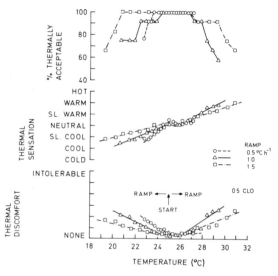

FIG. 12.11. Mean responses of subjects to slowly changing ambient temperatures. The subjects wore summer clothing with an insulation of 0.5 clo. (From Berglund, 1979.)

the higher the energy consumption (5 to 15% per °C). From the point of view of both comfort and energy it is thus essential that the buildings as well as the heating and air-conditioning systems be designed so that the occurrence of local cold discomfort is unlikely. For summer conditions the same applies for local warm discomfort.

On the other hand, local warming or cooling of the body may provide a means of energy conservation during winter or summer, respectively. This may be considered primarily under conditions where deviations from optimal comfort would be acceptable (Fanger, 1979).

13
Thermoregulatory Behaviour in Animals

13.1 Natural and Operant Behaviour

Thermoregulatory behaviour of animals under natural and experimental conditions has long been studied (for references see
Whittow, 1971; Hensel *et al.*, 1973; Bligh, 1973: Ingram and
Mount, 1975). A new approach to thermoregulatory behaviour was
started when Kundt *et al.* (1957a) observed behavioural changes as
well as autonomic responses during cooling the anterior hypothalamus in conscious cats. The animals curled up in quite the same way
they did in cold environments, and shivering as well as vasoconstriction occurred. This observation of the involvement of hypothalamic
temperature in the activation of thermoregulatory behaviour has been
confirmed in cats (Freeman and Davis, 1959) and in a variety of other
animals.

It was not until the last 20 years that operant behavioural methods
were extensively used as an approach to temperature regulation. This
technique was introduced by Weiss (1957) and consists in the training
of animals to operate a device in order to receive either warmth (e.g.
infrared radiation) or cold (e.g. a stream of cold air). When the
animals are exposed to a heat stress, their work rate for cold will
increase while a cold stress will increase their work rate for heat.

In recent years, operant behavioural methods have been combined
with artificial displacements of local temperatures of various thermosensitive sites in the body, and with a variety of non-thermal stimuli,
including the administration of drugs that influence temperature
regulation (for references see Satinoff and Hackett, 1977; Satinoff and
Hendersen, 1977; Satinoff, 1979). Of course, one has to take into
account that operant behaviour of animals in an artificial situation is

not necessarily the same as thermoregulatory behaviour under natural conditions. Furthermore, the possibility of unspecific behavioural responses, such as arousal, must be ruled out. Nevertheless, this approach has given valuable new insight into the process of temperature regulation.

13.2 Interaction of Thermal Drives

13.2.1 HYPOTHALAMUS AND SKIN

When dogs were allowed to choose between cold, neutral and warm environments, they usually preferred to be in a thermoneutral or warm environment. After the ingestion of hot water which raised core temperature by a few tenths of a degree, the dog selected the cold environment, but returned to the thermoneutral or warm environment when its core temperature had returned to its normal level (Cabanac et al., 1965, 1966). This behaviour was assumed to be caused by a rise in core temperature.

Cooling and warming the hypothalamus in various species, including rats (Satinoff, 1964, 1979; Carlisle, 1966, 1970; Corbit, 1969, 1970, 1973; Murgatroyd and Hardy, 1970), pigs (Baldwin and Ingram, 1967; Carlisle and Ingram, 1973b), sheep (Baldwin and Yates, 1977), baboons (Gale et al., 1970), and squirrel monkeys (Saimiri sciureus) Adair, 1969, 1970, 1974, 1976a,b, 1977a,b; Adair et al., 1970; Adair and Rawson, 1973, 1974) was followed by corrective behavioural responses.

With operant behavioural techniques it has been established that rats increase the frequency at which they press a bar to turn on a heat radiator during local cooling of the POAH region (Satinoff, 1964). The same hypothalamic cooling also induced shivering. It was concluded that behavioural and autonomic thermoregulatory responses are complementary, because the temperature rise was of the same order whether or not heat could be obtained behaviourally. An alternative explanation could be that the effect of external heating was small as compared with that of shivering and possibly non-shivering thermogenesis. In similar experiments in rats, Carlisle (1966) obtained a decrease in rate of bar pressing for external heat when the hypothalamus was locally heated.

These results were confirmed in baboons (Gale et al., 1970), which increased their rate of bar pressing for infrared heat when the POAH region was cooled. Pigs learned to press a pad with their snout to obtain a certain amount of radiant heat (Baldwin and Ingram, 1967,

1968b). When the hypothalamus was cooled, the bar-pressing rate for warmth increased, while it decreased when the hypothalamus was cooled. In addition, the behavioural response to hypothalamic temperature changes was dependent on ambient temperature, in that hypothalamic cooling was effective only at ambients of 0 to 25 °C, but not at 30 to 35 °C.

The interaction of hypothalamic and skin temperature in thermoregulatory behaviour has been studied in various species and with a variety of experimental setups. In rats and monkeys the interaction of peripheral and central temperature signals was measured quantitatively with respect to operant thermoregulatory behaviour. When rats are kept in various thermal environments, they operate heating and cooling devices in such a way that hypothalamic temperature remains fairly constant (Carlisle, 1968a,b). Corbit (1969, 1970, 1973) used an arrangement in which rats could control either their hypothalamic (T_h) or their skin temperature (T_s) while the corresponding temperature was displaced to various levels. The rate at which the animals worked for reductions in T_h was increased by increasing either T_s or T_h, which suggests that both thermal inputs contribute to the thermal discomfort. It can be assumed that thermal discomfort resulting from displacement of either T_h or T_s can be alleviated by changes in T_h as well as by changes in T_s. The data suggest a linear system with additive summation of both inputs for the response (R), according to the equation

$$R = a(T_h - T_{h,0}) + b(T_s - T_{s,0}) \qquad (22)$$

where a and b are slope constants expressing the relative importance of T_h and T_s in determination of the magnitude of the response.

A direct comparison of behavioural and autonomic temperature regulation was made in squirrel monkeys (Stitt et al., 1971). The monkey could influence his average skin temperature ($\bar{T}_s$) by choosing between warm and cold airstreams. Besides these behavioural responses, sweating and shivering were simultaneously recorded. When the temperature (T_h) of the medial preoptic hypothalamus was displaced to high values, the animal selected lower $\bar{T}_s$ values than normal, and at low T_h he selected higher $\bar{T}_s$ values than normal (Fig. 13.1). These results show a hyperbolic relation between T_h and $\bar{T}_s$ with regard to thermoregulatory behaviour, suggesting a multiplicative signal interaction. A similar relation was found for shivering and sweating. The controller for thermoregulatory behaviour thus seems identical in its characteristics to the controller for autonomic thermoregulatory responses. It can further be seen in Fig. 13.1 that thermo-regulatory behaviour occurs in a zone where neither shivering nor sweating is activated, in other words behavioural responses are more temperature-sensitive than shivering or sweating. Probably vasomotor responses are also highly sensitive to temperature changes but this type of control action has not been tested.

When monkeys were trained to regulate the temperature of a microheater implanted into the POAH region, they responded both to

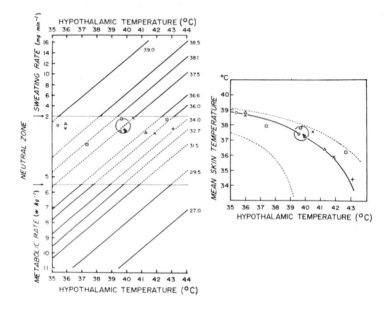

FIG. 13.1. Panel on right shows behaviourally preferred skin temperature at different values of hypothalamic temperature in squirrel monkeys. Panel on left shows these data points plotted on the physiological controller. The dotted lines on each graph encompass the thermoneutral zone where neither shivering nor sweating occurred. (From Stitt *et al.*, 1971.)

temperature changes of the microheater and of the air (Adair, 1977a); the warmer the heater and the warmer the ambient, the faster the monkeys responded to turn the heater off. In contrast to findings in the rat (Corbit and Ernits, 1974), preliminary results of a preference test have shown that a monkey trained to regulate both skin and POAH temperature prefers to cool the skin in response to POAH warming. In other experiments, attempts to train monkeys in a cool environment to warm the POAH behaviourally have resulted in failure, confirming earlier findings by Baldwin and Ingram (1967) in the pig. These results suggest that the drives from POAH and skin temperature for thermoregulatory behaviour, although interdependent, are not equivalent.

There are relatively few behavioural experiments in which the thermal stimuli were changed very slowly. An experiment of this type in the squirrel monkey is shown in Fig. 13.2. When the POAH temperature is linearly increased with a rate of about 2 °C h^{-1}, the decrease in the behaviourally selected air temperature occurs not

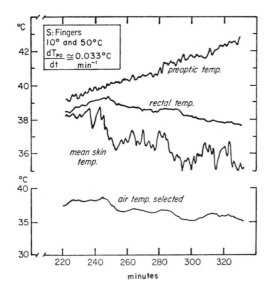

FIG. 13.2. Stepwise reduction of mean skin temperature and selected air temperature during slow warming of a squirrel monkey's preoptic area. T_{po}, preoptic temperature. (From Adair and Rawson, 1973.)

linearly but in steps, as can also be seen in the resulting skin temperature (Adair and Rawson, 1973). These steps occur also in autonomic metabolic and evaporative responses (cf. Fig. 8.4). The neural processes that transform a slow linear temperature change into a step-wise function are not yet known.

It has been shown in squirrel monkeys and sheep (Adair and Rawson, 1974; Adair, 1976b) that identical thermal stimuli applied bilaterally to the POAH region have a greater effect on thermoregulatory behaviour than has unilateral stimulation. On the contrary, when one side of the POAH region was cooled and the other side heated, the effects on behaviour were diminished or even abolished.

These findings suggest that both sides of the POAH region interact in an additive way with respect to thermoregulatory behaviour.

13.2.2 MEDULLA OBLONGATA

Operant behavioural responses could be elicited by heating and cooling the medulla oblongata in rats (Lipton, 1971, 1973). The responses were of the same order of magnitude as those induced by

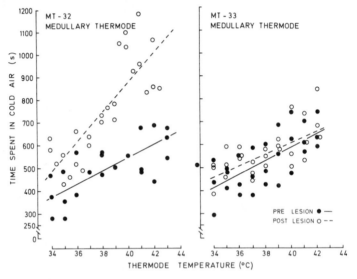

FIG. 13.3. Effects of forebrain lesions in rats on relation between medullary temperature and amount of time spent pressing a pedal for cold air. Destruction of PO/AH region (left) produced an enhancement of slope of behaviour as a function of medullary temperature. Injury posterior to PO/AH region (right) did not cause a change in behavioural response to medullary thermal stimulation. (From Lipton, 1973.)

temperature changes in the POAH region. When the latter was destroyed, thermoregulatory behaviour was intensified (Fig. 13.3), whereas destruction of the posterior hypothalamus did not change the behavioural responses to temperature changes of the medulla oblongata.

13.2.3 SPINAL CORD

Thermal sensitivity of the spinal cord was tested by behavioural methods in dogs and pigs. In the dog, spinal heating at warm ambients increased the rate at which a fan for cool air was operated (Cormarèche-Leydier and Cabanac, 1973). This rate was decreased when the ambient temperature was lowered. In contrast to the effects seen with spinal heating, no behavioural responses could be observed during spinal cooling.

In the pig, however, both spinal heating and cooling elicited behavioural responses (Carlisle and Ingram, 1973b). Cooling the

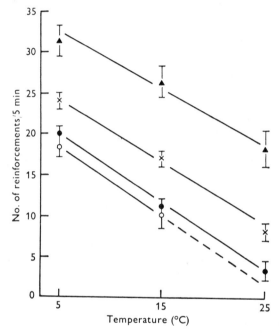

FIG. 13.4. Mean number of reinforcements in pigs per 5 min period averaged over 20 min before thermal stimulation(●), during cooling of the spinal cord to 20 °C(×), heating of the spinal cord to 43 °C (○), and cooling of the hypothalamus to 20 °C (▲) as a function of ambient temperature. Vertical lines indicate standard error of mean, and the dashed line is an extrapolation. (From Carlisle and Ingram, 1973b.)

spinal cord increased and warming decreased the rate of obtaining thermal reinforcements by operating infrared heat lamps (Fig. 13.4). The effect of spinal cooling lasted only for a brief period, an observation that was tentatively ascribed to the dynamic response of spinal cold receptors (cf. p. 109). As can be seen in Fig. 13.4, a decrease in ambient temperature considerably enhances the effect of spinal cooling, and so does local hypothalamic cooling.

13.2.4 HYPOTHALAMUS AND DEEP BODY SITES

When transient displacements (ΔT_h) of local preoptic temperatures (T_h) in squirrel monkeys are made around a neutral level of T_h, the changes behaviourally selected in air temperature (ΔR) are proportional to the amount of displacement (Adair, 1977b) (Fig. 13.5). The same general result was obtained when the displacements were made around a higher level of T_h, but in this case the function relating the behavioural adjustment of air temperature to the magnitude of

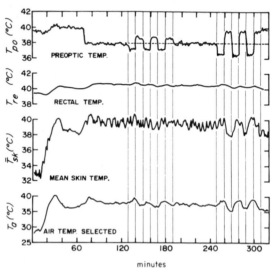

FIG. 13.5. Air temperature selected by a squirrel monkey and body temperature changes to preoptic temperature clamp at 37.8 °C and transient preoptic temperature displacements 0.7 and 1.5 °C above and below this level. Initial hour represents stabilization of behaviour and all measured parameters. (From Adair, 1977b.)

displacement $(\Delta R/\Delta T_h)$ was steeper, in other words, the gain was increased. The higher the clamped level of T_h, the higher was the gain.

Where can the thermal drives for this change of gain originate? Obviously not in the skin, because skin temperature (T_s) is under behavioural control of the animals and thus cannot govern preoptic thermosensitivity. However, an altered, but stable rectal temperature seems to be associated with the gain of the output of the effector system. Adair (1977b) assumes, therefore, that extrahypothalamic thermosensors elsewhere in the body may be involved. This relation can be expressed as

$$\bar{T}_s \text{ selected} = \beta \ (T_h - T_{h,0}) \qquad (23)$$

where $T_{h,0}$ represents the set temperature. β is defined by the behavioural hypothalamic sensitivity $(\Delta R/\Delta T_h)$ and depends on the temperature of the extrahypothalamic body core.

The view that intraabdominal thermoreceptors may be involved in thermoregulatory behaviour is further supported by experiments in which heating devices were chronically implanted into the abdomen of conscious animals. Marked changes in thermoregulatory behaviour

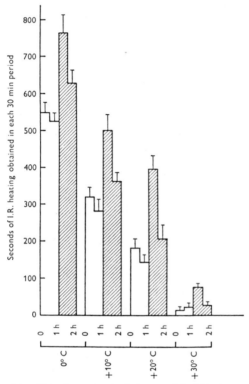

FIG. 13.6. Effect of loading the rumen of sheep with 1 litre water at 0 to 1 °C at ambient temperatures of 0, 10, 20 and 30 °C, on the duration of infrared heating obtained by shorn sheep. The unhatched bars indicate the duration of heating obtained in the 1 h control period preceding loading, and the hatched bars indicate the duration of heating in the 1 h period following loading. Mean results from three sheep. Vertical bars indicate standard error of mean. (From Baldwin, 1975.)

of squirrel monkeys were observed after local thermal stimulation of the colon (Adair, 1971). Cooling caused an increase in operant cutaneous warming that was linearly related to the degree of cooling. Since hypothalamic temperature and abdominal skin temperature did not change during the initial phase of colonic stimulation, the behavioural responses were ascribed to thermosensitive structures somewhere in the abdomen.

Loading the rumen of shorn sheep with 1 litre water of 0 to 1 °C caused a a marked increase in the rate of operating an infrared heater (Baldwin, 1975). This thermoregulatory behaviour was also strongly dependent on ambient temperature, as seen in Fig. 13.6. When 2 litres water at the same temperature were given, the behavioural response was approximately twice as large as with 1 litre. These responses may possibly be due to the excitation of ruminal cold receptors (cf. p. 115), although it cannot be excluded with certainty that hypothalamic or other internal thermoreceptors are also involved. These findings agree with experiments by Bykov (1959) and his collaborators who have shown by conditioning techniques that dogs could distinguish between water at 26 or 36 °C when applied to the stomach via a fistula. They also demonstrated that dogs could distinguish between water at 7 or 28 °C when it was introduced into the intestines.

Other indications for the involvement of extrahypothalamic deep body thermosensors for behavioural responses can be derived from experiments in sheep (Baldwin and Yates, 1977). When the anterior hypothalamus was locally cooled at ambient temperatures of 10 °C, the animals turned on an infrared heater more frequently, but after a period of 2 h the rate was reduced and the reduction coincided with a rise in deep body temperature of 0.75 °C. Conversely, hypothalamic warming resulted in a reduction in the rate of operating the heaters during the first 85 min, but then the use of heaters increased and this coincided with a fall of about 0.75 °C in deep body temperature. However, it cannot be decided whether the thermal drive comes from spinal cord thermosensors or other deep body thermoreceptors. Cooling the brain by intracarotid injections of cold saline had the same effect on thermoregulatory behaviour as had local hypothalamic cooling. Even small temperature changes within the normal range of hypothalamic temperatures were sufficient to elicit marked behavioural responses.

13.3 Autonomic Thermoregulation and Behaviour

If animals have the choice, do they prefer to regulate their body temperature autonomically or behaviourally? Cabanac et al., (1970) observed that some dogs responded inefficiently for infrared heat in cold environments or convective cooling in warm environments, supplementing the behaviour by shivering or panting. Other dogs did not shiver or pant, utilizing thermoregulatory behaviour to spare the

autonomic responses. The reasons for this different type of behaviour
are not yet known.

Under certain circumstances, such as during exercise, autonomic
regulation may dominate. However, much of the time homoiotherms
exhibit a complex set of behavioural responses aimed at providing a

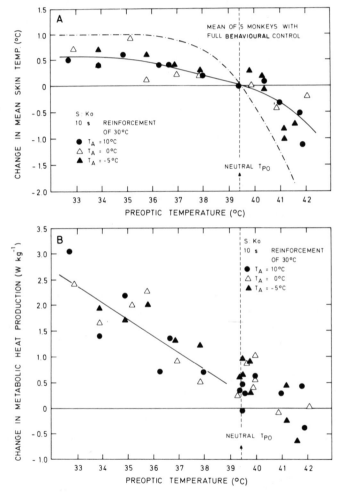

Fig. 13.7. A: Change in mean skin temperature of one monkey with limited
behavioural control during last 30 min of 1-h preoptic temperature (T_{po}) clamp
relative to mean skin temperature during the 30 min preceding each clamp. T_A,
Ambient temperature. Data are compared with a comparable function for five other
monkeys that have full behavioural control over the environment. B: Change in
metabolic heat production relative to level during 30 °C T_A baseline period as a
function of preoptic temperature (T_{po}). (From Adair, 1976a.)

hospitable microclimate and their dependence upon autonomic responses is minimal. The same is generally true for humans. Hardy
(1972) proposed that only if behavioural thermoregulation is made
difficult or unacceptable in some way, will the organism use its
autonomic abilities to establish thermal balance. Many investigators
(Adair *et al.*, 1970; Corbit, 1970; Hardy, 1972; Bligh, 1973) have
pointed out that when an endotherm has complete behavioural
control over his environmental temperature, there is no necessity for
him to resort to such responses as shivering or panting in order to
achieve balance. With vasomotor control that is continuously active
around the neutral range it may be different.

When the local preoptic temperature of a monkey is raised or
lowered relative to the neutral level, the monkey responds by selecting
a lower or higher air temperature. Hence, even under conditions of
local preoptic temperature change, autonomic responses do not occur
because behaviour alone can effectively and rapidly bring about a
compensatory change in skin and core temperature. Only when the
behavioural responses are limited or difficult to perform, the animal
will exhibit autonomic thermoregulatory responses. This was shown
in squirrel monkeys by restricting the availability of warm air for
behavioural compensation of hypothalamic cooling (Adair, 1976a). In
this case the average skin temperature could not be raised to its
compensatory level and an increase in metabolic heat production
occurred (Fig. 13.7). The data demonstrate a strong preference for the
behavioural response mode over the autonomic. In other experiments
(Adair and Wright, 1976) the effort to perform the behavioural
response was gradually increased. Even in this case the animals
continued to respond as long as they were physically able and
relinquished thermoregulatory behaviour only gradually although
their autonomic temperature regulation was available to them. Behavioural inadequacy was compensated for by appropriate autonomic
responses to regulate the body temperature.

It would be interesting to observe the interaction of thermoregulatory behaviour and autonomic temperature regulation under natural
conditions or in situations where other motivations become important. A monkey restrained in a chair and having nothing else to do,
might well prefer thermoregulatory behaviour, whereas under certain
natural circumstances, e.g. when seeking food or avoiding a predator,
autonomic temperature regulation may be predominant.

Using implanted thermodes that allowed separate thermal stimulation of the POAH region and the posterior hypothalamus in squirrel
monkeys, Adair (1974) investigated the question whether autonomic

responses and thermoregulatory behaviour can as well be elicited from the posterior hypothalamus. Cooling the POAH region resulted in cutaneous vasoconstriction and increase in rectal temperature, while warming caused vasodilatation and a subsequent fall in rectal temperature. However, in the posterior hypothalamus the same temperature changes were nearly ineffective.

In contrast to these findings, thermal stimulation of the posterior hypothalamus was highly effective in initiating thermoregulatory behaviour. Figure 13.8 shows that the behaviourally selected air temperature following thermal stimulation of the posterior hypothalamus was nearly the same as when stimulating the POAH region. Additional experiments were conducted to investigate the interaction of anterior and posterior hypothalamus with respect to thermoregulatory behaviour (Fig. 13.9). The function of both areas seems to be

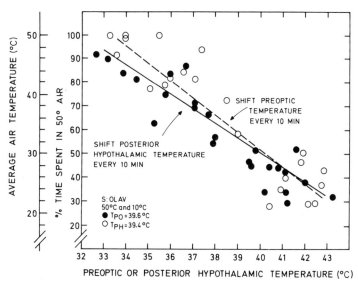

FIG. 13.8. Behavioural selection of air temperature by a squirrel monkey. Percentage of each 10-min interval 50 °C air was chosen (and resulting average air temperature) is plotted as a function of preoptic (open circles, dashed line) or posterior hypothalamic (closed circles, solid line) temperature. Preoptic or posterior hypothalamic temperatures were shifted every 10 min from normal levels to a temperature indicated on the abscissa. (From Adair, 1974.)

connected in an additive way, in that clamping the temperature of the posterior hypothalamus at low or high levels, respectively, will lead to parallel shifts in the behavioural response curve to preoptic temperatures.

The results show a clear uncoupling of autonomic and beh
thermoregulatory responses. While autonomic reactions are
restricted to thermal stimulation of the POAH region, therm
tory behaviour can be elicited by temperature changes
preoptic and posterior hypothalamus. It has been shown tha
of the preoptic area, while impairing autonomic thermore
responses, appear to leave thermoregulatory behaviour intact
(1968) has observed that rats with rostral lesions were
effectively regulate against heat but increased the rate of be
heat-escape responses. Similar results were obtained in col
ments (Carlisle, 1969; Satinoff and Rutstein, 1970). The m
tent consequence of preoptic and anterior lesions was an increas

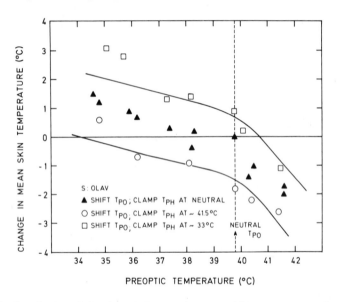

Fig. 13.9. Steady-state behavioural changes in mean skin temperature of a squirrel
monkey as a function of preoptic temperature when posterior hypothalamic temperature was simultaneously clamped at neutral (solid triangles), clamped at 41.5 °C
(open circles), or clamped at 33 °C (open squares). Solid lines represent predicted
relationships based on additivity of action. (From Adair, 1974.)

rate of response for radiant heat reinforcement; in spite of this, rectal
temperature was often lower than normal. The results show that an
impaired autonomic regulation can be compensated by operant
behaviour. How are these behavioural responses mediated? In this
case the posterior hypothalamus may be an important source of
thermal drive for thermoregulatory behaviour.

14
Displacement of Set Point

14.1 Assessment of Set Point

The set point of a proportional control system has been defined as that value of the controlled variable at which the control action is zero. In order to determine the set point, it is thus necessary to measure the control actions. In a multiple-loop system the control action can be zero at various combinations of temperatures. Therefore, simultaneous temperature measurements in various representative sites of the body, such as hypothalamus, spinal cord, and skin, are required. For practical purposes, the weighted mean body temperature $(\bar{T}_b)$ may be suitable (cf. p. 11).

As already mentioned, the set temperatures of various autonomic effector responses, such as shivering or panting, are not necessarily identical (Table 14.I). They can also be influenced separately and

TABLE 14.I

Mean thresholds of thermoregulatory control actions in humans

Shivering threshold $\bar{T}_b$	Vasomotor threshold $\bar{T}_b$	Sweating threshold $\bar{T}_b$	Remarks	Author
35.2 °C[a]		36.6 °C[a]		Baum et al. (1976)
36.7 °C[b]		36.75 °C[b]		
36.13 °C	36.09 °C[c]	36.11 °C	Cold bath 28 °C	Cabanac and Massonet (1977)
37.17 °C	35.16 °C[d]	37.46 °C	Women	Cunningham et al.
36.61 °C	34.70 °C[d]	37.20 °C	Men	(1978)

T_b, Weighted mean body temperature.
[a] Subjects with area body mass quotient A_D/m [10^{-4} m^2 kg^{-1}] = 220.
[b] Subjects with A_D/m = 300.
[c] Threshold for heat loss from hand.
[d] Threshold for mean tissue conductance.

independently. Furthermore, the set point for thermoregulatory behaviour can differ from that of autonomic responses. According to the
state of adaptation and other circumstances, there may be a smaller or
larger gap between the thresholds for metabolic heat production and
evaporative heat dissipation ("inter-threshold zone", cf. Chapter 15);
in contrast, some authors have found that heat production and
evaporative heat dissipation may even slightly overlap (Hardy, 1961;
Jacobson and Squires, 1970; Cabanac, 1970; Jessen, 1976, 1977).

In practice one can find accessible sites whose temperature can be
viewed as representative of that of the whole body, provided that
comparable conditions can be adhered to. The rectal temperature or
the sublingual temperature is often used for this purpose in medical
practice. Fluctuations and deviations of these representative temper-
tures from an empirically arrived at "normal value" may be due to
the following conditions: (1) Thermal stress leads to a load error; (2)
the control system is overtaxed—this refers to hypo- and hyperthermia; (3) the standard conditions of measurement are no longer valid,
the temperature field of the body having changed; (4) the set point of
the thermoregulatory system has been displaced. Usually a displacement can be assumed if the representative internal temperature shows
a deviation while the effector processes are minimal and the temperature field of the body remains practically unchanged.

From the concept of a multiple-input system it follows that any
temperature in the body that gives rise to a feedback signal is part of
the controlled variable. If this view is accepted, then it follows that
temperature signals *per se* do not shift the set point. Thus we should
consider a displacement of the set point primarily in the cases in which
non-thermal factors are involved.

14.2 Normal Set Point Displacements

14.2.1 CIRCADIAN VARIATION

The body temperature of man and other homeotherms fluctuates in a
daily rhythm; the amplitude of this fluctuation is 0.7 to 1.5 °C in adult
man (Aschoff, 1970; Conroy and Mills, 1970; Hildebrandt, 1974;
Scheving *et al.*, 1974; Palmer, 1975). It has long been suspected that
neither sleep nor food intake nor muscular activity is essential for the
maintenance of a 24-h rhythm in body temperature, and evidence has
accumulated that the rhythm persists if the above factors are ruled out
(Hildebrandt, 1974). Under resting conditions, spontaneous variations of heat production are small compared with those of peripheral

vascular tone. It is therefore evident that the nycthemeral variations in body temperature are brought about mainly by changes in heat dissipation (for references see Hildebrandt, 1974).

Since the 24-h rhythm of body temperature is present in a temperature range where the control actions are near zero, the only theoretical possibility to explain this variation is a shift in set point. Any change in body temperature caused by other factors, such as a change in effector responsiveness, would easily be compensated by feedback mechanisms.

Cabanac et al. (1976) have applied behavioural methods to assess the set point fluctuations during the nycthemeral cycle. Human subjects were submerged in a water bath, the temperature of which could rapidly be changed from 30 to 40 °C, thereby changing the oesophageal temperature (T_{oe}) from hypothermic to hyperthermic levels at relatively constant skin temperatures of 40 °C. The set point was defined as that value of T_{oe} at which a series of cold and warm stimuli applied to the hand were rated neither pleasant nor unpleasant. At higher T_{oe}, cold stimuli were pleasant and warm stimuli unpleasant, whereas the opposite occurred at lower T_{oe} (cf. Fig. 12.2).

Figure 14.1 shows the average set point assessed by behavioural methods, as well as T_{oe} at various times of day. There is a pronounced nycthemeral fluctuation, the rising phase starting in the early morning and the falling phase in the early afternoon. The results suggest a close correlation between autonomic and behavioural regulation.

Temperature is only one of many physiological parameters subject to circadian rhythms (for references see Pittendrigh, 1974; Palmer, 1975; Aschoff, 1979). The periodic temperature fluctuations are due to circadian oscillations of hitherto unknown origin that are synchronized with the earth's rotation (local time) by means of synchronizers (time-givers, entraining agents, or cues). Usually the most effective synchronizer is the lighting regimen but any other periodic meteorological or ecological event might act in this way. In mammals and birds it is easy to cause a phase shift in diurnal rhythm in regard to local time by shifting the phase of the synchronizer; in contrast, the frequency of the period can only be influenced to a limited extent. A phase shift in all synchronizers is known to cause a corresponding phase shift in the temperature curve in man as well, most completely when there is a change in local time during transmeridional travel. A matter of high practical importance is the time course of adaptation to the new rhythm. For a phase shift of about 12 h, the time required for synchronization to the new phase is 7 to 12 days (Halberg, 1969; Scheving et al., 1974; Palmer, 1975).

Human subjects isolated from their environment for a span of up to six months continue to exhibit a circadian rhythm of internal temperature desynchronized from 24-h local time; the "free-running" periods vary between 24.4 and 26 h but in some cases the variability is much greater. During the free-running period, a desynchronization

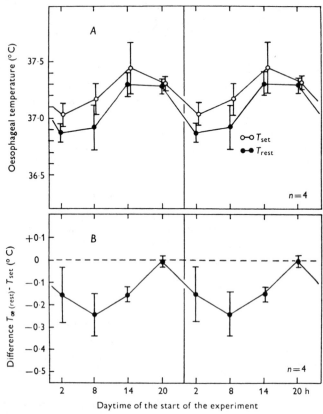

Fig. 14.1. A: Nyctohemeral cycling of mean set point (T_{set}) in human subjects, estimated from behavioural results (open circles) compared to oesophageal temperature (T_{oe}) measured at the end of the control 30 min resting period (filled circles). B: Average differences between $T_{oe(rest)}$ at the end of the resting period and T_{set}. T_{oe}, oesophageal temperature. (From Cabanac et al., 1976.)

between temperature rhythm and other biological rhythms, such as sleep and wakefulness, can occur (Colin et al., 1968; Aschoff, 1970).

In squirrel monkeys even a desynchronization between various components of thermoregulation was observed during free-running periods (Fuller et al., 1979), and thermoregulation was impaired in an environment without circadian time cues (Fuller et al., 1978). Thus effective thermoregulation appears to require the precise synchronization of the internal time keeping system.

In rodents (Saleh *et al.*, 1977; Stephan and Nunez, 1977) and squirrel monkeys (Fuller *et al.*, 1977), lesions of the suprachiasmatic nuclei have been found to abolish or impair the diurnal pattern of body temperature, the mean level being approximately the same as the 24-h mean temperature of intact animals.

14.2.2 SLEEP

Besides the sleep-independent displacements of set point due to the 24-h cycle, other changes in temperature regulation have been

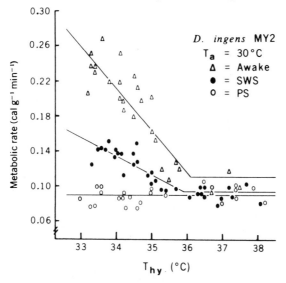

FIG. 14.2. Relationship between metabolic rate and hypothalamic temperature (T_{hy}) during three states of arousal for one kangaroo rat. Each symbol represents the average metabolic rate during an uninterrupted period of wakefulness. SWS, slow wave sleep; PS, fast wave or paradoxical sleep: T_{hy}, manipulated hypothalamic temperature; T_a, ambient temperature. (From Glotzbach and Heller, 1976.)

observed that are specifically correlated to sleep. Baker and Hayward (1967) suggested that thermoregulation in the rabbit is inhibited during certain phases of sleep that are characterized by fast wave activity in the EEG and by rapid eye movements (REM). During the periods of fast wave sleep in cats, cold shivering as well as panting and cutaneous vasomotion were absent and reappeared during the slow wave and spindle phases of sleep (Parmeggiani and Rabini, 1967;

Parmeggiani and Sabattini, 1972; Parmeggiani *et al.*, 1977; Parmeggiani, 1977).

The hypothesis that during fast wave sleep thermoregulation is inhibited or suspended was further supported by artificial heating of the preoptic-anterior hypothalamic (POAH) region in cats. During fast wave sleep, the thresholds for panting as well as for cutaneous vasodilatation were shifted to higher hypothalamic temperatures (Parmeggiani *et al.*, 1977). Local cooling of the hypothalamus to 33 °C in kangaroo rats (*Dipodomys*) did not cause an increase in metabolic rate, whereas during slow wave sleep and in awake animals the

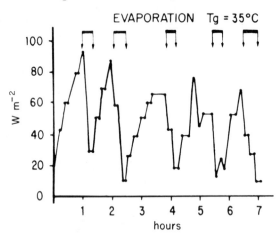

Fig. 14.3. Changes of evaporative rate during nocturnal sleep of a representative subject. Black horizontal bars with arrows indicate REM periods. (From Henane *et al.*, 1977a.)

threshold for metabolic increase was at hypothalamic temperatures near 36 °C (Glotzbach and Heller, 1976). During slow wave sleep there was also a decreased slope of the metabolic response compared to that in awake animals (Fig. 14.2).

There are also significant changes in the spontaneous activity of hypothalamic neurons during the phases of fast wave sleep (Findlay and Hayward, 1969; Parmeggiani and Franzini, 1971, 1973). Although the thermoregulatory significance of these changes has not been established, Parmeggiani (1977) suggests that the pattern of hypothalamic activity during fast wave sleep may reflect a new mode of operation of thermoregulatory structures.

The changes in thermoregulation during fast wave sleep can be described as a widening of the interthreshold zone between heat production and heat dissipation, thereby rendering the animals poikilothermic within a larger band of body temperatures (Parmeggiani, 1977). Since this change is particularly pronounced in small mammals, it was thought that the poikolothermic phase during fast wave sleep has a sparing function for metabolic energy comparable to hibernation (Glotzbach and Heller, 1976).

Sweating responses in man are suppressed during fast wave or REM sleep, even in warm environments of 37 to 39 °C (Shapiro *et al.*, 1974; Henane *et al.*, 1977a, 1979). As Fig. 14.3 shows, there is an almost complete inhibition of sweating during the REM phases and a subsequent rise in average skin temperature, but the effects on body temperature are relatively small. Thus the thermoregulatory changes occurring during fast wave sleep seem to be less important for humans than they may be for small mammals.

14.2.3 MENSTRUAL CYCLE

The menstrual cycle evokes a periodic fluctuation in body temperature. In humans the preovulatory rectal temperature is about 0.5 °C lower than the temperature in the postovulatory phase. According to the criteria mentioned above, these temperature fluctuations have to be ascribed to displacements of the set point. This assumption is supported by the investigation of thermal comfort in relation to the preovulatory and postovulatory phase of the menstrual cycle. When female subjects assessed thermal comfort or discomfort elicited by local stimulation of the hand (cf. Fig. 12.2), the judgments change in such a way during the menstrual cycle as to indicate a shift in set point (Cunningham and Cabanac, 1971).

14.3 Set Point and Physical Work

The body temperature in homeotherms increases during physical work and activity. After equilibrium is attained the temperature level depends on the intensity of the work but is relatively independent of the ambient temperature. This shift in body temperature does not necessarily indicate a displacement of the set point; it can just as well be explained as follows: the heat produced by physical work creates a load error. Even if this load error is relatively small, the central temperature rises to a high level as a consequence of the low skin temperature caused by evaporative heat loss. Skin temperature during work, even in a warm environment, may be lower than it is in the zone of thermal neutrality under resting conditions (Nielsen, 1969). A similar situation can be achieved without evaporative heat loss when the skin is artificially cooled by a water-circulated garment (Webb, 1970).

An additional factor during severe work in humans may be the cooling of the brain by heat exchange from face and respiratory tract (Cabanac and Caputa, 1979). In a multiple-loop control system this would allow an even higher increase in deep trunk temperature without a shift in set point. For example, marathon runners can attain a deep trunk temperature of 41.9 °C without clinical signs of heat stress (Maron et al., 1977), which means that their thermoregulatory capacity is not overtaxed and their temperature is regulated.

Although the rectal temperature may be taken as an approximate measure of the regulated temperature under resting conditions, it is no longer valid during work. As a result of altered conditions of heat formation and of evaporative heat loss, the temperature field must then necessarily be different and an increase in rectal temperature cannot be taken as proof for a set-point displacement.

When the same amount of heat was generated by active work and by passive heating through diathermia, the rectal temperature increased in both cases to about the same level and appeared, within limits, to be independent of the environmental temperature. At identical mean skin temperatures, the increase in thermal conductance and sweat rate was also practically equal for active and passive heating (Nielsen, 1969). This remarkable similarity in thermoregulatory response to both conditions strongly suggests that the increase in body temperature is not due to a shift in set point but to the factors mentioned above.

There are other indications that the set temperature is not elevated during physical work. It has been found that the metabolic rate remains almost unchanged during light work in a cold environment, even when the core temperature decreases (Kitzing et al., 1968), which is hardly compatible with the assumption of an upward shift in set point. Kitzing and his colleagues (Kitzing et al., 1971) have compared the input and output of human temperature regulation at widely varying ambient temperatures (−5 to +40 °C) and workloads (0 to 294 W) in the steady state. If sweat rate, thermal conductance and cold-induced metabolic increase were predicted from oesophageal and skin temperatures, the same equation (Behling et al., 1971) was valid for all combinations of ambient temperatures and workloads. It was therefore concluded that during steady state of rest and exercise, the only systematic input variables to the system of thermoregulation are well characterized by mean skin and core temperature and that no shift in set point occurs during physical work.

Further evidence of a constancy in set temperature during work can be derived from observations of thermal comfort. Cabanac *et al.* (1971) allowed the subjects to adjust the temperature of one hand to his own choice of comfort. When the subject exercised for a short time

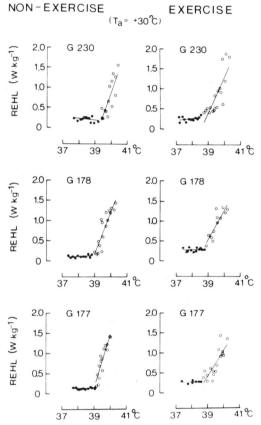

Fig. 14.4. Comparison of the relationship between respiratory evaporative heat loss (REHL) and body core temperature (T_a) during rest and exercise (3 km h⁻¹, 5° gradient) in three goats (G230, G178, and G177). Each symbol represents the average REHL during last 15 min of a 20-min period in which T_a was held constant with an intravascular heat exchanger. Open and closed circles represent data included in calculating indicated linear regression lines. (From Mercer and Jessen, 1979.)

with a high workload, no change in hand preference temperature was seen, suggesting that non-thermal inputs did not enter the system to displace the set point.

Various experimental findings suggest that the set temperature for evaporative heat loss may even be lowered at the onset of work. In dogs exercising in warm air, there was an immediate increase in panting at the beginning of the exercise and an immediate decrease at the end (Hammel, 1968). The same holds true for thermoregulatory salivation in the dog (Hammel and Sharp, 1971). These findings may be interpreted as an initial decrease of the set temperature with the onset of work.

Similar observations have been made for sweating in man at the onset of exercise (Nielsen, 1969; Crockford et al., 1971; Nadel and Stolwijk, 1971; Nadel et al., 1971b; Saltin and Gagge, 1971; Tam et al., 1978), and also during uphill, level, and downhill running. With the same expenditure, the sweat rate at higher running rates was greater than it was at lower rates, although, by virtue of the experimental arrangement, the internal temperature was 0.2 to 0.3 °C lower at the higher sweat rate and the skin temperature was the same (Nielsen, 1969). The mechanism of this enhancement of sweating during work still remains obscure. Some authors have assumed that the sweating drive originates from thermoreceptors in the muscles; others suggest that, when work is begun, sweat secretion may be stimulated by sympathetic excitation based on non-thermal factors (Nadel and Stolwijk, 1971; Nadel et al., 1971b). For exercise in a state of dehydration, changes in the osmotic and/or ionic environment of the hypothalamus have been assumed to be involved in the elevation of body temperature (Senay, 1979).

Measurements of respiratory evaporative heat dissipation in dogs (Clough and Jessen, 1974) and goats (Mercer and Jessen, 1979) with clamped hypothalamic and spinal cord or hypothalamic and body core temperature, respectively, have shown that the response curves to certain combinations of thermal drives undergo only minor changes during exercise (Fig. 14.4). This indicates that the set point was practically not shifted by non-thermal influences from physical exercise.

14.4. Fever

This section deals only with changes in temperature regulation during fever. For more details about the important question of the biological and clinical significance of fever, the reader is advised to see, for example, Lwoff (1969); Eggers (1971); Schmidt (1975a,b); Klastersky and Kass (1978); Haahr and Mogensen (1978); Banet (1979); Kluger (1979a,b).

14.4.1 FEVER AND SET POINT

According to the previous definition, it must be assumed that, during fever, the set temperature of the thermoregulatory system is elevated.

This can already be concluded from simple observations during a typical acute attack of fever. At the beginning, physiological responses to cold appear, and the core temperature rises to a new level at which thermal neutrality is achieved (Fig. 14.5). As fever abates, regulatory processes appear that are opposed to excess warming; as a result the core temperature falls to the normal level again. Fever is distinguished from hyperthermia principally by the fact that the regulatory processes are hardly burdened at all in fever, whereas in hyperthermia they are working at the limit of their capacity.

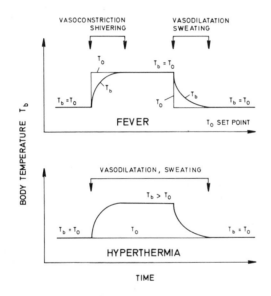

FIG. 14.5. Fever and hyperthermia. T_b, body temperature; T_0, set point. For further explanation see text.

In addition to these observations there is ample experimental evidence for the theory of a displacement of the set point in fever. If, in animals, the hypothalamus is locally heated during the onset of fever, the febrile responses and the increase in body temperature can be suppressed (Eisenman, 1972). By this method the load error caused by the higher set temperature is diminished. A steady-state heat load applied to rabbits leads to virtually the same input/output relation of the thermoregulatory system under normal conditions and during steady-state fever induced by i.v. infusion of leucocyte pyrogen (Cranston et al., 1976a). In goats there was no difference in the shivering reaction to central cooling between normothermic and fevered animals (Al Hachim and Frens, 1975).

Likewise, the input/output relation of control actions of human temperature regulation was unaffected by the administration of pyrogen or by the administration of salicylates after pyrogen, the only difference being a higher temperature level at which the responses occurred (Cooper *et al.*, 1964a; Rosendorff and Cranston, 1968; Cooper, 1972). When subjects were immersed in water baths of various temperature, they chose different hand temperatures as pleasant (cf. Fig. 12.2). During fever these behavioural response curves were shifted to a higher level (Fig. 14.6) but their slope was only slightly changed (Cabanac and Massonet, 1974).

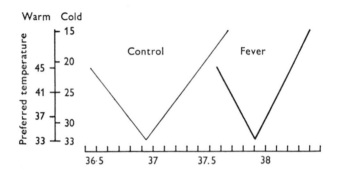

FIG. 14.6. Preferred hand temperature (warm, cold) as a function of oesophageal temperature (T_{oe}) in a human subject during fever and in a control subject. The skin temperature was kept constant in a water bath of 40 °C. (From Cabanac and Massonet, 1974.)

As Mitchell *et al.* (1970) have suggested, fever might be a change of gain rather than a shift in set point. This assumption is at variance with numerous findings demonstrating a shift in set point during fever without appreciable changes of the input/output relations of temperature regulation. The only exception so far has been found in rabbits where the metabolic response to local temperature changes in the preoptic-anterior hypothalamic (POAH) region showed a decreased sensitivity to warming (Eisenman, 1972, 1974b). However, when the mean skin temperature was kept constant, there was only a shift in set point without a change in POAH sensitivity (Stitt *et al.*, 1974). Possibly the decreased sensitivity to hypothalamic heating may be due to the higher skin temperature during the plateau phase of fever.

14.4.2 SITES OF PYROGEN ACTION

Various agents which produce fever (exogenous pyrogens, ExP), such as bacteria, viruses and endotoxins, stimulate the organism to produce pyrogenic substances called endogenous pyrogen (EnP). EnP is mainly derived from leucocytes (leucocyte pyrogen) but can also be produced by other tissues, e.g. the reticuloendothelial system (for references see Hellon, 1975; Siegert *et al.*, 1976; Siegert, 1977a,b; Dinarello *et al.*, 1977; Dinarello and Wolff, 1978; Kluger, 1979a,b; Lipton, 1980).

According to Dinarello (1980), human EnP produced by peripheral blood leucocytes is a polypeptide consisting of two fractions with molecular weights of 15 000 and 40 000. The larger molecule appears to be a trimer of the smaller one. Squirrel monkeys developed fever in response to relatively small intravenous and intracerebral doses of highly purified human leucocytic pyrogen with a molecular weight of 15 000 (Lipton *et al.*, 1979).

Recently R. Siegert (personal communication) succeeded in isolating an endogenous pyrogen. This EnP from granulocytes in the peritoneal exudate of rabbits, could be characterized chemically and immunologically; it was a homogenous protein fraction with a molecular weight of 14 300 and an extremely high pyrogenic activity (50 ng per kg body weight). A second protein fraction with a molecular weight of 70 000 had a much lower pyrogenic activity. This fraction was assumed to be a precursor of the smaller molecule.

There is no evidence so far that pyrogens have a direct action on peripheral structures. Cold receptors in the tongue of the rabbit were not influenced by administration of ExP (Frens, 1971), nor was a thermal reflex elicited from cutaneous warm receptors affected in man (Bryce-Smith *et al.*, 1959). In a patient with complete transection of the spinal cord at the level of C6, EnP did not cause vasoconstriction in the paralysed hand, although the spinal autonomic reflexes were intact (Cooper *et al.*, 1964b). Neither did direct injection of EnP into the subarachnoidal space of the spinal cord in rabbits elicit any fever response (Rosendorff and Mooney, 1971). These findings indicate that EnP does not act on the nervous system below the level of C6 or on other peripheral tissues.

Fever can readily be produced in various species of homeotherms by injecting EnP directly into the POAH region (Cooper *et al.*, 1967; Jackson, 1967; Repin and Kratskin, 1967; Rosendorff and Mooney, 1971). A similar effect can be observed after local injection of ExP into the POAH (Villablanca and Myers, 1965; Repin and Kratskin, 1967;

Myers *et al.*, 1971a,b, 1974; Lipton and Fossler, 1974) but the latency is longer than that seen after administration of EnP (Cooper *et al.*, 1967; Jackson, 1967). It is not clear whether this effect of ExP involves the local production of EnP at the site of the injection. Other sites of the brain, except the brain stem reticular formation (Rosendorff and Mooney, 1971) have been found to be irresponsive to local microinjections of pyrogens.

When antipyretics (e.g. salicylate) are given intragastrically, the fever caused by intrahypothalamic injection of pyrogens is suppressed

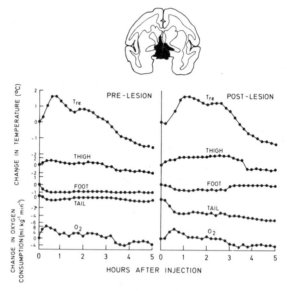

Fig. 14.7. Febrile response to intravenous endotoxin ($100\,\mu g\,kg^{-1}$) before and after destruction of POAH region in a monkey. (From Lipton and Trzcinka, 1976.)

(Myers *et al.*, 1971b, 1974). In just the same area of the POAH, in which microinjections have their greatest effect, microinjections of sodium salicylate cause an immediate fall in the fever produced by a constant infusion of EnP (Cranston and Rawlins, 1972). Thus at least part of the antipyretic effect of salicylate is locally mediated in this area.

Based on these findings, the theory was put forward that the anterior hypothalamus is the sensitive target for EnP, although the question remains open how the large molecules on EnP can reach this brain area. However, there is evidence that the hypothalamus is not the only site of pyrogen action. For example, Veale and Cooper (1975) reported that, after removal of the entire POAH of rabbits, an i.v. injection of EnP still produced fever of similar magnitude to that found in control rabbits. Similar results have been reported in goats (Andersson *et al.*, 1965). In squirrel monkeys (Fig. 14.7) the capacity to develop fever after intravenous and intracerebroventricular injection of ExP, i.v. injection of EnP, and intracerebroventricular injection of prostaglandin E_1 remained unchanged when the POAH region was destroyed (Lipton and Trzcinka, 1976). These results suggest that in the brain there is either multiple central representation of fever control or an inherent capacity to develop sensitivity to pyrogens and to produce coordinated febrile responses.

14.4.3 PUTATIVE MEDIATORS OF PYROGEN ACTION

The question as to whether monoamines, such as noradrenaline (NA) and 5-hydroxytryptamine (5-HT) are involved in febrile responses cannot satisfactorily be answered as yet. Depletion of brain monoamines by various agents led to conflicting results in various experiments and species (for references see Hellon, 1975). In rabbits, the responses to pyrogens did not change when 5-HT was depleted and cold defence reactions were impaired by pretreatment with *p*-chlorophenylamine (Borsock *et al.*, 1977). Hellon (1975) suggests that monoamines act mainly as transmitters in afferent pathways and that their possible involvement in fever might be a consequence rather than a cause. This view is further supported by the fact that fever caused by intrahypothalamic administration of prostaglandin E_1 in monkeys is not substantially influenced by intrahypothalamic injection of 5-HT or NA (Myers and Waller, 1976).

In recent years, the prostaglandins (PG) of the E series have extensively been discussed as possible mediators of pyrogen action (for reference see Hellon, 1975; Siegert, 1977a,b; Kluger, 1979a,b). This view is based on the following facts: (1) Small doses of PGE_1 or E_2 injected into the third ventricle or microinjected into the anterior hypothalamus of various mammals cause similar changes in temperature regulation as seen during fever (Milton and Wendlandt, 1971; Feldberg and Saxena, 1971b,c; Hales *et al.*, 1973; Stitt, 1973; Stitt *et al.*, 1974; Veale and Cooper, 1974, 1975; Crawshaw and Stitt, 1975; Cooper *et al.*, 1976b; Veale and Wishaw, 1976; Simpson *et al.*, 1977; Barney and Elizondo, 1978). The thermoregulatory changes following intrahypothalamic injection of PGE_1 can be described as an upward displacement of set point without a change in central thermosensitivity (Stitt *et al.*, 1974). During PGE_1 fever, the elevated temperature is maintained regardless of ambient temperature or whether behavioural or autonomic means are utilized (Crawshaw and Stitt, 1975). (2) After a pyrogen is given, the PG content in the fluid of the third ventricle and the cisterna in cats rises to a level 10 to 20 times higher than the initial value. After intraperitoneal injections of the antipyretics indomethacin, paracetamol or salicylate, the temperature falls and at the same time the concentration of PG shows a sharp decrease towards the normal level (Feldberg and Gupta, 1973; Feldberg et al., 1973). (3) Antipyretics such as salicylate and indomethacin reduce the synthesis of PG from its precursor arachidonic acid, by inhibition of the PG synthetase enzyme (Vane, 1971; Smith and Willis, 1971; Ferreira et al., 1971; Flower and Vane, 1973).

From these findings the hypothesis was derived that EnP induces the formation of PG in the hypothalamus and that PG acts in some way upon hypothalamic neurons involved in the displacement of set point. However, there are several findings which are hardly compatible with this hypothesis: (a) Marked dissociations have been found between PG and pyrogens in their ability to induce fever in certain species, or at some time during their life cycle, and in animals with chronic POAH lesions (for references see Kluger, 1979a). (b) Single-unit studies of hypothalamic thermosensitive neurons have revealed fundamental differences between the effects of PG and pyrogens (p. 216). (c) When a steady-state fever was caused in rabbits by i.v. infusion of EnP, appropriate doses of salicylate abolished the increase of PGE in the cerebrospinal fluid without affecting the fever (Cranston et al., 1975). (d) When fever was produced in rabbits by the injection of EnP in a lateral cerebral ventricle, the latency, rate of rise and magnitude of the fever was unaffected by the simultaneous intraventricular injection of two prostaglandin antagonists, SC 19220 and HR 546. Both antagonists effectively attenuated the fever caused by the intraventricular injection of PGE_2 (Cranston et al., 1976b).

This evidence is not consistent with the hypothesis that PGE is the principal mediator of fever. Laburn et al. (1977) found that arachidonic acid leads to the production of at least two pyrogenic substances—prostaglandins and either prostaglandin endoperoxide or thromboxanes. An intraventricular injection of the sodium salt of arachidonic acid resulted in a dose-dependent fever in rabbits. This fever was blocked by indomethacin, whereas the prostaglandin antagonist, SC 19220, was ineffective. Laburn et al. (1977) concluded that some breakdown products of arachidonic acid other than prostaglandins might be pyrogenic.

To summarize, more experiments are necessary before any putative mediator can be assigned a role in pathogen-induced fevers. The same holds for cyclic $3',5'$-adenosine monophosphate (cyclic AMP) which has also been suggested as a putative mediator of fever (for references see Siegert, 1977a; Kluger, 1979a).

14.4.4 NEURONAL ACTIVITY

Microelectrode recordings have shown that thermosensitive neurons in the POAH region and in the midbrain reticular formation can be influenced by ExP and EnP. In the rabbit (Cabanac et al., 1968; Wit and Wang, 1968b) as well as in the cat (Eisenman, 1969, 1970, 1972), there was clear evidence of a change in sensitivity of warm- and cold-sensitive neurons, whereas the activity of thermally insensitive neurons was not influenced by ExP. According to Eisenman (1969), high Q_{10} units classified as primary warm sensors showed a rotation in their frequency/temperature curves around a point near 38 °C after the injection of pyrogen (Fig. 14.8). This means that their firing rate remains unchanged during normal body temperature and thus cannot be the source of the modified drive to the effectors. On the other hand, units whose characteristics suggest that they were interneurons in the regulatory pathways showed both a change in sensitivity and in firing rate after pyrogen administration. Warm-sensitive cells decreased their sensitivity and firing rates, while cool-sensitive ones increased their firing rates. These responses are appropriate for the decrease in heat loss and increase in heat production that would be seen in an unanaesthetized animal during the rising phase of a pyrogen-induced fever.

Thermosensitive neurons in the mesencephalic reticular formation were also influenced by ExP. All cold-sensitive units increased their firing-rate level but did not change their temperature sensitivity, while the discharge of warm-sensitive neurons disappeared after administration of pyrogen. Thermally insensitive neurons remained unaffected (Nakayama and Hori, 1973).

When a bacterial endotoxin is given in urethane-treated cats, a febrile reaction and a depression of the activity of single warm-sensitive units in the POAH area are seen. As Wit and Wang (1968b) have demonstrated, this depression can be reversed by the administration of acetylsalicylate. More recent experiments with leucocyte pyrogen have confirmed these results (Schoener and Wang, 1975). After microinjection of EnP into the POAH region, warm-sensitive neurons were depressed and cold-sensitive neurons enhanced in their activity. Micro-injection of sodium salicylate without pyrogen pre-

treatment caused no significant modification of the thermal responsiveness of POAH neurons in most cases. However, when salicylate
was administered after pyrogen, it uniformly antagonized the pyretic
effect causing a return of the discharge to the control rate. This is
further evidence that EnP and antipyretics, at least in part, are acting
on the same site.

Microiontophoresis of PGE_1 into the POAH region of conscious
rabbits influenced a constant proportion of 8 to 10% of warm-
sensitive, cold-sensitive and thermally insensitive units, the response

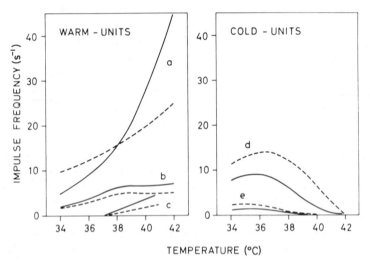

Fig. 14.8. Effect of pyrogen on hypothalamic neurons. Solid lines: before pyrogen
administration. Broken lines: after pyrogen administration. (Curves a, b, and d from
Eisenman, 1972; c and e from Cabanac *et al.*, 1968.)

being invariably one of mild facilitation. Thus the results cannot
predicate any simple neuronal basis for the action of PGE_1 in
producing fever (Stitt and Hardy, 1975).

Our picture of neural events during fever has not essentially been
clarified by the single-unit studies. On the contrary, new questions
have arisen: How can the discrepancy between the change of gain
seen in thermosensitive neurons and the shift in set point of thermoregulatory responses without a marked change of gain be explained?
Furthermore, there are fundamental differences between the effect of
pyrogens and PGE on single unit activity, findings that do not support
the assumption that PG is a mediator of pyrogen action.

14.5 Cations and Set Point

Myers and Veale (1970, 1971) have observed in cats that marked changes in body temperature could be produced by changing the Na^+/Ca^{2+} ratio of fluids during an intrahypothalamic perfusion. An increase in the ratio caused a rise in body temperature and a decrease of Na^+/Ca^{2+} a fall in temperature. In monkeys the results were confirmed with intraventricular and intrahypothalamic perfusions (Myers et al., 1971c; Myers and Yaksh, 1971). Microperfusions of isolated areas of the diencephalon with solution containing 11.0 to 34.0 mM excess Na^+ or 11.3 to 47.9 mM Ca^{2+} altered the body temperature of unanaesthetized monkeys only when the sites of perfusion were located in the mammillary region of the posterior hypothalamus. After the prolonged temperature rise produced by excess Na^+ or the deep fall in temperature evoked by excess Ca^{2+}, the monkey responded to intragastric application of cold water with shivering and vasoconstriction in the ear pinna and to hot water with vasodilatation at any level of body temperature. Chelation of Ca^{2+} by perfusing the posterior hypothalamus with ethylene glucol bis-β-aminoethyl ether-N, N, N', N'-tetraacetic acid (EGTA) caused a rapid elevation in temperature that could be abolished immediately by perfusing a solution of excess Ca^{2+} at the same diencephalic site. The hypothermia caused by excess Ca^{2+} in unanaesthetized cats was not abolished by the calcium antagonists verapamil and xylocaine (Metcalf and Myers, 1976).

As far as can be concluded from the experiments, there is some indication that the set point was actually displaced. For example, shivering was absent at any level of the lowered body temperature but could immediately be elicited by intragastric cooling. On the other hand, this procedure also caused shivering at body temperatures well above 40 °C, although, upon intragastric cooling, the temperature fell to only 39 °C. However, the temperature effects of excess Ca^{2+} or Na^+ in the ventricular system could be reversed by raising the ambient temperature. In dogs kept at 21 °C, high Ca^{2+} or Na^+ caused hypo- or hyperthermia, respectively, but at 28 to 33 °C these actions were reversed, which would not be the case if the ions caused a shift in the set point of temperature regulation (Dhumal and Gulati, 1973). In exercising rats, the elevated colonic temperature could be lowered by hypothalamic perfusion of excess calcium (Gisolfi et al., 1976). When unanaesthetized monkeys exercised strenuously, the effect of labelled $45\ Ca^{2+}$ increased markedly in the push–pull perfusate of the diencephalon, and similar changes were seen during external cooling (Gisolfi et al., 1977).

It would be highly desirable to combine the perfusion experiments with elaborate thermophysiological studies, thus providing more relevant information about the function of the stimulated structures and the nature of the observed changes. As to the importance of the experimental findings for natural temperature regulation, additional research is necessary.

It was further reported that in cats the Na^+/Ca^{2+} ratio in the cerebrospinal fluid changed with the rise in temperature produced by an endotoxin. This change was associated with an increased efflux of labelled Ca^{2+} and a decreased efflux of labelled Na^+ into the cerebroventricular fluid, whereas the effect was reversed by administration of an antipyretic, acetaminophen (Myers, 1976). The assumption that the hypothalamic Na^+/Ca^{2+} ratio may play an important role in the change of set point during fever (Myers et al., 1976a,b) is tempting in connection with the finding that Ca^{2+} enhances the activity of peripheral warm receptors and inhibits that of cold receptors (p. 61). But the evidence is not sufficient to establish a similar central mechanism, and in humans no significant change of the Na^+/Ca^{2+} ratio in cerebrospinal fluid has been found so far during fever (Nielsen et al., 1973).

15
Long-term Thermal Adaptation

15.1 General Aspects

When homeotherms are exposed to chronic deviations of their thermal environment, they develop adaptive modifications of already existing regulatory processes. Such adaptations occur in the single individual, their time course ranging between hours and months, sometimes even years. The slowest events are genetic adaptations, involving several generations and time amounting perhaps to millions of years. Genetic adaptations are not dealt with in the following.

Long-term thermal adaptation requires an additional feedback loop that induces chronic changes in the properties of the regulatory system. The essential factor acting through this loop is not temperature *per se* but time, be it a single-step change in environmental temperature, a steadily changing thermal stimulus, or an intermittent stimulation. As in acute thermoregulation, there are also behavioural responses in long-term thermal adaptation. In this respect, man's civilizational and technological achievements can be considered as the most specific and efficient behavioural adaptations to the thermal environment.

Stabilization of a physiological function in a changed environment is possible when other functions deviate from their initial value as long as the chronic stimulus ("stressor"; Adolph, 1964) is being maintained. These compensating functions are called "adaptates". The time course of stabilization thus depends on the temporal development of adaptates.

Adaptations are not always single events reaching a new steady state aperiodically after the environment has changed. They may involve a sequence of adaptive responses with various time constants. The early adaptates are often overshooting and transient reactions. When exposure is prolonged, the initial adaptates may decline, while

219

more stable adaptive responses will appear. An example is the sequence of adaptates in homeotherms during prolonged cold exposure: (i) shivering thermogenesis, (ii) non-shivering thermogenesis in connection with transitory changes of the endocrine system, (iii) perhaps non-shivering thermogenesis with normalized endocrine functions, and (iv) trophic changes in the integument. As a general rule, one can say that the early adaptates are mainly functional and not as specific and economical as are the late responses of a structural

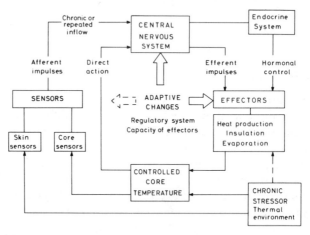

Fig. 15.1. Adaptive modifications of the thermoregulatory system. For further explanation see text.

or morphological type. Thus the biological freedom and independence of an organism from its environment will gradually be improved with increasing duration of the exposure, provided that the strength of stimulus does not exceed a critical level.

The quality of adaptates depends not only on the strength and duration of the stressor, but also on its time structure (e.g. constant, slowly changing, continuous, intermittent). Adaptation in the presence of a chronic thermal stimulus and loss of adaptation (deadaptation) in the absence of further exposure do not have the same kinetics, the disappearance usually being slower than the development of adaptates.

Adaptive modifications that are beneficial for the organism under extreme thermal conditions may also be the result of stressors other than temperature. This phenomenon is called "positive cross adaptation". For example, increased physical fitness through exercise may lead to an increased cold resistance. On the contrary, there are examples of "negative cross adaptation". Thus adaptation to low levels of oxygen in the inspired air has been shown to decrease cold resistance (for references see Hensel *et al.*, 1973).

In spite of an enormous mass of data regarding physiological and biochemical changes during prolonged exposure to various temperatures (for references see Chaffee and Roberts, 1971; Bligh, 1973; Hensel *et al.*, 1973; Wyndham, 1973; Ingram and Mount, 1975; Le Blanc, 1975; Janský and Musacchia, 1976; Lind, 1977; Brück, 1978c; Wang and Hudson, 1978), our knowledge of the adaptive mechanisms is extremely limited. Present evidence supports the view that, in the last analysis, thermal adaptation can be traced back to modifications in the central nervous system. In addition, some peripheral mechanisms may be involved.

Long-term changes in the thermal environment can reach the CNS either directly, by deviation of core temperature, or indirectly, via thermoreceptors and afferent pathways (Fig. 15.1). In addition there remains the theoretical possibility that substances formed in other tissues under the influence of chronic temperature stimuli might act on the CNS. Changes in nervous mechanisms during long-term thermal adaptation may occur at any level of the thermoregulatory system. They can involve the acquisition of new responses as well as the qualitative or quantitative change of existing responses. In contrast to acute regulation, the neural induction of hormonal responses plays an important role in long-term thermal adaptation (Collins and Weiner, 1968; Smith and Horwitz, 1969; Chaffee and Roberts, 1971; Hensel *et al.*, 1973; Eastman *et al.*, 1974).

Table 15.I summarizes various functional and morphological modifications during long-term adaptation to thermal environments.

TABLE 15.I

Physiological modifications during long-term thermal adaptation

Functional		Morphological
Regulatory system	Capacity of effectors	
Threshold deviation	Non-shivering thermo-genesis	Body size and shape
Change of gain	Peripheral blood flow	Brown adipose tissue
Thermal behaviour	Sweat rate	Capillarization
	Respiratory rate	Fur thickness
		Fat blubber
		Sweat glands

Changes in body size and shape can occur during the ontogenetic development of animals raised at different ambient temperatures. (Weaver and Ingram, 1969; Ingram, 1977; Heath and Ingram, 1978). Similar modifications are also known as genetic ecogeographical variations of homeothermic species (for references see Hensel *et al.*, 1973). Adaptive modifications in thermal insulation such as pelage growth are not only induced by temperature changes. Rather, the growth of fur depends on periodic events that are synchronized with the seasons mainly by the length of daytime. This has the advantage that a higher thermal insulation is attained before the cold season starts (for references see Hensel *et al.*, 1973).

15.2 Habituation to Local Thermal Stimuli

Immersion of the human hand in water of 4 °C causes an increase in arterial blood pressure and heart rate. When the cold stimuli are repeated several times with intervals of 1 min, a marked decrease of the blood pressure response is seen. This "short-term habituation" does not occur when the time intervals are extended to 5 min (Strempel and Tändler, 1976).

An explanation of these phenomena may be derived from the response of cutaneous cold receptors (Schäfer and Hensel, 1975). Rapid cooling of the skin to 4 °C leads to a high dynamic overshoot of the receptor discharge (cf. Fig. 4.1). If the skin is rewarmed, the initial state of the receptor discharge is not reached after 1 min. Thus the dynamic overshoot caused by the following cold stimulus is much smaller than the previous one. In fact, the course of the blood pressure response and that of the dynamic overshoot of cold receptors as function of repeated cold stimuli and time intervals is strikingly similar.

If series of cold stimuli applied to the hand are repeated for 10 to 30 days, there is also a marked long-term habituation of the blood pressure response (Fig. 15.2). At the same time, a marked adaptation of local cold pain is observed (Glaser, 1966; Le Blanc, 1975; Strempel, 1978). These adaptive changes are strictly confined to the hand exposed to the repeated cold stimuli (Glaser, 1966; Le Blanc, 1975). While the blood pressure response is stressor-unspecific, the process of habituation is specific; for example, if habituation to cold stimulation is achieved, immersion of the hand into hot water will still elicit a rise in blood pressure (Glaser, 1966). A spatial specificity of local habituation has also been demonstrated (Eide, 1971, 1976).

From the experience of everyday life it is known that people who are used to working with bare hands in the cold are able to stand local cooling that is presumed to be very painful. A number of investigators have studied various ethnic groups that are naturally cold exposed, special professional groups, such as fishermen, fish filleters, and groups of indoor workers before and after local cold exposure applied under laboratory conditions. Less or no pain was experienced during cold exposure of hands or fingers by all groups accustomed to

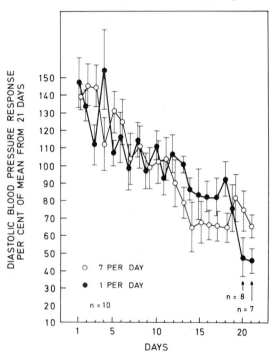

FIG. 15.2. Average responses of diastolic blood pressure in human subjects during repeated immersion of the hand in cold water. Brackets indicate standard error of mean. (From Hildebrandt, 1977.)

exposing their hands to cold (Le Blanc, 1975; Le Blanc *et al.*, 1975). In addition, there are some differences in the blood flow to the extremities adapted to local cold stimuli (p. 235).

Long-term habituation to local cold stimuli seems to be due to changes in the central nervous system. These mechanisms have been studied in the rat (Glaser, 1966). The tail was submerged in water of 4 °C and the increase of heart rate was observed. This response was shown to diminish to about half the initial level after 10 days when the procedure was repeated every day for short periods. Habituation was impeded when lesions were made in the frontal areas of the brain. Furthermore, electrical stimulation of the frontal regions was found to facilitate habituation.

15.3 Changes in Thermal Afferents

A much-discussed problem is the possibility of adaptive changes in the afferent input from cutaneous thermoreceptors during long-term

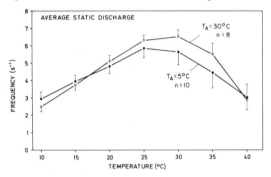

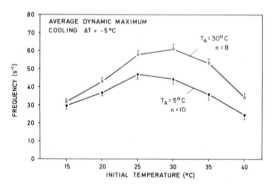

Fig. 15.3 Average static discharge of cold receptors in the cat nose and maximum dynamic response to a 5 °C cooling step as function of initial temperature. The animals were adapted to ambient temperatures (T_A) of 30 and 5 °C, respectively, for a period of 4 years. (From Hensel and Schäfer, unpublished.)

exposure to various ambient temperatures. When cats were continuously kept for 2 months at room temperatures of 35 and 5 °C, respectively, no considerable change was seen in the static and dynamic characteristics of cold receptors in the nasal area (Hensel and Banet, 1978). However, more pronounced changes occurred when the cats were living at 30 and 5 °C, respectively, for about 4 years (Hensel and Schäfer, 1979). As shown by Fig. 15.3, the average static frequencies between 40 and 17 °C were somewhat lower in the cold-adapted group, the respective values for both groups at 30 °C being 5.6 and 6.5 s^{-1}. At 10 °C, however, the frequency was higher in the cold-adapted animals. The average dynamic maxima were considerably lower in the cold-adapted group throughout the whole temperature range, the values at 30 °C being 44 and 61 s^{-1}, respectively ($P = 0.001$). In addition, the maxima of the static and dynamic curves were shifted from 30 to 25 °C in the cold-adapted group.

It cannot yet be decided whether these long-term changes are due to alterations of the cold receptors or of the thermal properties of the skin. In any case, the static thermal afferents at 18 °C skin temperature are the same for cold- and warm-adapted cats. It can therefore be concluded that the peripheral thermal drive at low ambient temperatures is practically not influenced by adaptive changes of the input from cutaneous cold receptors.

The weight of a clipped standard area of fur was about 35 % higher in the cold-adapted group, and at 0 °C ambient temperature the metabolic rate of the cold-adapted animals was 18% lower than that of the warm-adapted ones. It seems unlikely that these adaptive changes were maintained by a chronic fall in core temperature, since at ambients of 5 °C the cold-adapted cats had rectal temperatures that were about the same as those of warm-adapted cats at 25 °C. For shorter periods of cold adaptation this may be different; in the beginning of cold exposure the rectal temperature may decrease by more than 2 °C and thus induce adaptive responses by way of central cooling. It may therefore be assumed that the thermal drives for the described adaptive modifications of long duration originate mainly in the periphery.

15.4 Threshold Deviations of Thermoregulatory Responses

Deviations in threshold temperatures or set points for various thermoregulatory responses are important processes involved in long-term thermal adaptation. During prolonged cold exposure of various species of small mammals, such as rats and guinea pigs, shivering is gradually reduced while non-shivering thermogenesis increases considerably (for references see Hensel et al., 1973; Janský and Musacchia, 1976). This change is mainly due to a shift in the shivering

threshold to lower mean body temperatures, as shown in Fig. 15.4
(Brück and Wünnenberg, 1970). Thus the already existing capacity of
non-shivering thermogenesis will be activated, while, at the same
time, shivering will be suppressed by the "meshed" control system
(cf. p. 157 and 245). In cold-adapted guinea pigs, this shift from
shivering to non-shivering thermogenesis can also be demonstrated
when the metabolism is increased by administration of pyrogens

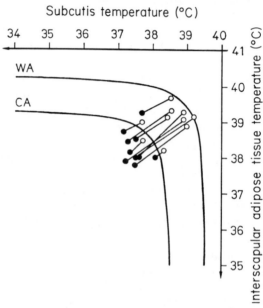

FIG. 15.4. Lowering of the threshold temperature for the onset of shivering during
i.m. infusion of noradrenaline (○, before; ●, after) in warm-adapted guinea pigs
compared with the threshold hyperbolas for cold-adapted (CA) and warm-adapted
(WA) animals. (From Zeisberger *et al.*, 1977.)

(Blatteis, 1976a). In humans, non-shivering thermogenesis is of major
importance only during infancy (see Chapter 16).

   The first evidence of adaptive threshold deviations in humans came
from studies in central Australian aborigines (Hammel *et al.*, 1969).
These natives overcome cold stress 1) by an increased functional heat
insulation through extremely reduced peripheral blood flow and 2) by
tolerating some drop in body temperature without shivering or
increased heat production. A striking feature of this "insulative-
hypothermic" adaptation (Hammel, 1964) is the ability to tolerate
low foot temperatures, which obviously includes habituation to cold

pain (p. 222). Similar adaptations have been found in Bushmen of the Kalahari Desert and in nomadic Lapps (Hammel, 1964). The questions are inherited or acquired during an individual's life.

A decrease in threshold temperature for shivering and cold-induced heat production can also be obtained in humans by repeated cold exposure. Such deviations have been found in the women pearl divers (ama) of Korea, who dive several hours a day in cold water (Hong,

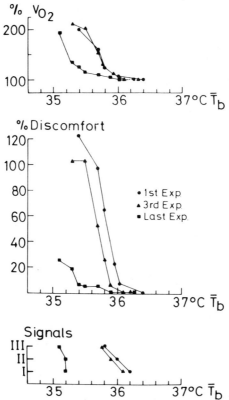

FIG. 15.5. Summary of a study in a human subject who was cold-exposed on six different days. The data from three cold exposures are represented. T_b, mean weighted body temperature. Signals: I, "cool"; II, "cold"; III, "very cold"; (●) 1st, (▲) 3rd, (■) final cold-exposure. Oxygen uptake, V_{O_2}, "100%" = initial resting level. (From Brück *et al.*, 1976.)

1963, 1973), in sojourners in the Antarctic (Bodey, 1978) as well as in subjects repeatedly exposed to cold in the laboratory (Davis, 1961; Brück *et al.*, 1976). An example is shown in Fig. 15.5. Wearing a bathing suit, the subjects were exposed four to seven times within 2 weeks to ambient temperatures between +5 and −5 °C, each exposure lasting 1 h. Metabolic reactions and shivering threshold were shifted to a lower weighted mean body temperature as well as a lower oesophageal temperature. This modification in the thermoregulatory system was linked with a reduction in thermal discomfort and general cold sensation. The latter changes will further be discussed on p. 237.

It should be emphasized that the hypothermic type of adaptation is not always seen when humans are exposed to cold environments. In

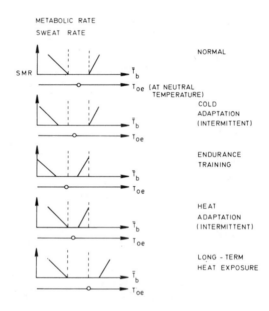

FIG. 15.6 Survey on the possible deviations of the thresholds for thermoregulatory metabolic and sweating responses in man. Open circles indicate set point shift. (From Brück, 1978c.)

the study of Brück *et al.* (1976) only two-thirds of the subjects showed this response. Moreover, in field studies where subjects lived in the cold for several weeks, a "metabolic" type of adaptation was observed, with high blood flow and temperatures in the extremities at the expense of an increased metabolism (Scholander *et al.*, 1958). In this case, physical exercise was thought to be the decisive factor in producing adaptive changes (for references see Hensel *et al.*, 1973).

When newborn guinea pigs were continuously cold exposed for several weeks, the thresholds for both shivering and heat polypnoea were shifted to a mean body temperature about 1 °C lower than that of controls. However, intermittent alternative exposure to heat and cold led to a downward shift in shivering threshold without a corresponding shift in threshold for heat polypnoea (Brück et al., 1971). This widening of the "interthreshold zone" allows an economical temperature regulation in oscillating thermal environments, in that the organism can use its heat capacity as a buffer without activating physiological reactions against heat and cold.

In human subjects, decrease in shivering threshold with the sweating threshold unchanged has been observed after intermittent cold exposure as a basis for "hypothermic" or "tolerance" adaptation

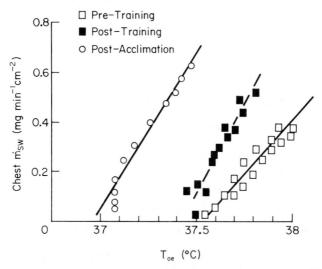

FIG. 15.7. Chest sweat rate plotted versus oesophageal temperature (T_{oe}) for one subject exercising at an ambient temperature of 25 °C. Open squares = pre-exercise training data; filled squares = postexercise training data; circles = postheat-acclimation data. (From Roberts et al., 1977.)

(Fig. 15.6). In this case, temperature regulation is not improved; on the contrary, the precision of control is reduced but it allows the subject to tolerate somewhat cooler environments without metabolic waste and emotional arousal (Brück, 1976; Brück et al., 1976). A downward shift in both shivering and sweating threshold of about 1 °C weighted mean body temperature (Fig. 15.6) was found in long-distance runners (Baum et al., 1976). These findings were confirmed in another series of experiments in long-distance runners and swimmers (Dressendorfer et al., 1977). This cross adaptation may be advantageous as it prolongs the time until a limiting high body temperature is reached.

In contrast to intermittent cold exposure, intermittent heat exposure of humans leads to a lowering of the sweating threshold (Fig.
15.6) and thus to a narrowing of the interthreshold zone (Henane and
Valatx, 1973; Gonzalez et al., 1974; Schwennicke and Brück, 1976;
Roberts et al., 1977; Brück, 1978c). In this case the body temperature
is more effectively defended against an upward deviation. A similar
decrease of the sweating threshold can also be found after endurance
training (Wyndham, 1973; Nadel et al., 1974; Henane et al., 1977b;
Roberts et al., 1977; Brück, 1978c).

During repeated heat exposure, no change of gain in the relationship beteen sweat rate and body temperature was seen (Henane
and Valatx, 1973; Gonzalez et al., 1974). However, the gain of the
sweating response (Fig. 15.7) increased during endurance training
(Henane et al., 1977b; Roberts et al., 1977). Thus the gain constant for
sweating was $168 \, W \, m^{-2} \, °C^{-1}$ in control subjects, 222 in long-
distance swimmers and 269 in cross-country skiers (Henane et al.,
1977b).

In contrast to the results of repeated heat exposure, it has been
found that sojourners in a hot climate develop tolerance adaptation
against heat (Fig. 15.6), that is, their mean body temperature is
higher after adaptation by more than 0.5 °C, but no sweating is
induced at this temperature (Raynaud et al., 1976).

15.5 Neurophysiological Correlates of Thermoadaptive Modifications

According to our present knowledge, changes in the characteristics of
cutaneous thermoreceptors, if present at all, cannot account for the
adaptive shift in threshold of thermoregulatory responses. No results
are as yet available which demonstrate a conceivable adaptive change
in internal thermosensitivity.

A concept for the interpretation of the threshold changes in guinea
pigs was based on the observation that intrahypothalamic injection of
noradrenaline (NA) may cause changes of threshold temperatures for
thermoregulatory reactions which resemble those found in the course
of cold adaptation (Zeisberger and Brück, 1971b; Brück, 1978c). In
the hypothalamus, a thermosensitive preoptic area and a more
caudally situated NA-sensitive area can be distinguished. Microinjection of NA into this area under thermoneutral conditions increases
shivering or non-shivering thermogenesis, or both. These metabolic
reactions can be suppressed by increasing skin and/or internal body
temperature (Zeisberger and Brück, 1973; Zeisberger et al., 1977). It
can be concluded that the threshold for the thermogenetic reaction is

shifted to a higher level by intrahypothalamic injection of NA (Fig. 15.8). Phentolamine, an adrenergic alpha-blocking agent, has the reverse effect when given intrahypothalamically: shivering threshold is decreased in warm-adapted animals to values normally found in cold-adapted and newborn guinea pigs (Zeisberger and Brück, 1976; Zeisberger, 1978). As can be seen in Fig. 15.8, the shift in threshold is combined with a change of gain of the thermogenetic response as function of temperature.

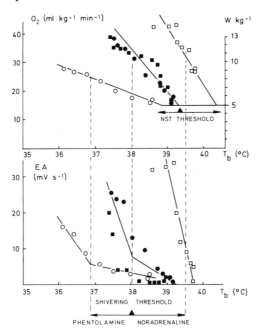

FIG. 15.8. Non-shivering thermogenesis (O_2 consumption) and shivering (muscular electrical activity, EA) in cold-adapted guinea pig as function of weighted mean body temperature $(\bar{T}_b)$ under normal conditions (●, ■), after intrahypothalamic administration of $1\,\mu g$ noradrenaline (□) and $20\,\mu g$ phentolamine (○). (From Zeisberger, 1978.)

These results have been summarized on the basis of a neuronal model (Brück, 1978c) shown in Fig. 15.9. The cold defence reactions are initiated and sustained by the action of peripheral cold receptors and inhibited by deep body warm receptors; the action of warm receptors is mediated by the inhibitory interneurons 1 and 2. An ascending catecholaminergic pathway is postulated (cf. p. 79) which makes synaptic contact with the interneurons 1 and 2. The action of

this pathway would then be seen in inhibiting the interneurons 1 and
2, thereby suppressing the inhibitory activity on the effector neurons 4
and 5; thus shivering and non-shivering thermogenesis would be
activated by this disinhibition. Phentolamine would have the opposite
effect by blocking the catecholaminergic pathway, thereby enhancing
the inhibitory activity of the interneurons 1 and 2 and thus suppressing shivering and non-shivering thermogenesis.

Our knowledge of possible adaptive changes in the activity of
central neurons involved in thermoreception and temperature regulation is extremely limited. Only recently the first evidence of a changed

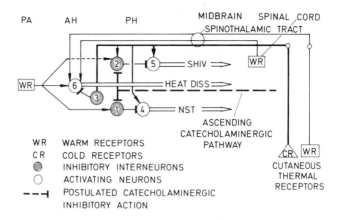

Fig. 15.9. A neuronal model of the central integrative parts of the temperature
control system derived from experiments in guinea pigs. PA, Preoptic area; AH,
anterior hypothalamus; PH, posterior hypothalamus; WR, warm receptors; CR, cold
receptors; NST, non-shivering thermogenesis. For further explanation see text. (From
Brück, 1978c.)

response characteristics of neurons in the central thermoreceptive
pathway has been found during long-term cold adaptation (Werner,
Schingnitz and Hensel, unpublished). When rats were exposed for
5 weeks to ambients of +3 °C, warm-reactive thalamic and midbrain neuronal populations decreased their threshold temperature for
scrotal warming by 1 to 2 °C (Fig. 15.10). According to our present
knowledge, changes in the properties of scrotal thermoreceptors can
be ruled out. These findings would mean that long-term cold adaptation not only decreases the threshold temperature for metabolic
responses, but may also decrease the threshold for heat defence
reactions. Similar shifts in threshold for cold- and heat-defence

reactions have been found in humans after endurance training (cf. Fig. 15.6).

It is not known as to whether thermoadaptive processes are induced by central or peripheral temperature changes, or by both. It is assumed that heat adaptation in humans may be elicited both by repeated central and peripheral heating (Marcus, 1972; Henane and Valatx, 1973). Some indirect evidence that an increased internal body temperature plays an important role can be seen in the fact that subjects tolerating a rise in central temperature develop a higher degree of heat adaptation than subjects who respond to hyperthermia

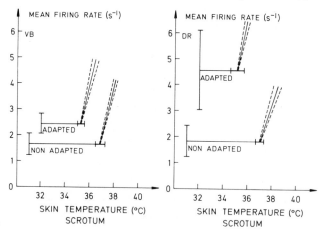

FIG. 15.10. Simplified response curve of thalamic (VB) and midbrain (DR) neurons to scrotal warming in cold-adapted rats and non-adapted controls. Mean values and standard errors of means of relevant parameters. (From Werner et al., 1980.)

with strong heat-defence reactions, thereby keeping their internal temperature at a lower level (Hednane and Valatx, 1973).

Repeated local cooling of the preoptic area of unanaesthetized rats for about 10 to 15 days was found to increase the capacity of non-shivering thermogenesis to a value about two times that of the control animals, as established by a standard injection of noradrenaline (Banet and Hensel, 1976b). Similar effects were seen after repeated cooling of the spinal cord (Banet and Hensel, 1976c). However, in contrast to animals adapted by external cooling, repeated hypothalamic or spinal cord cooling did not lead to a higher cold resistance of the animals.

Prolonged and intermittent cooling of the spinal cord in rats has been found to increase the gain of the temperature control system, as shown by a steeper increase of the metabolic response as a function of decreasing ambient temperature (Fig. 15.11), whereas prolonged and intermittent cold exposure of the animals led to a parallel shift of the metabolic response to higher values without a change of gain (Banet *et al.*, 1978b).

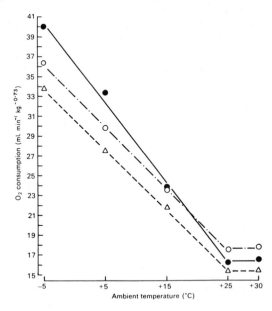

Fɪɢ. 15.11. Average oxygen consumption at various ambient temperatures in rats after prolonged intermittent spinal cord cooling (●), after prolonged intermittent exposure to cold ambients (○) and in controls (△). (From Banet *et al.*, 1978.)

15.6 Changes in Effector Systems

Only a general outline of thermoadaptive changes in effector systems will be given, as far as these modifications are relevant for thermal adaptation in humans. For more detailed information, the references on p. 155 may be consulted.

15.5.1 ᴍᴇᴛᴀʙᴏʟɪsᴍ

The cold-adaptive shift from shivering to non-shivering thermogenesis is combined with an increase in the metabolic response to a standard cold stress (for references see Hensel *et al.*, 1973; Heldmaier, 1974a). This increased metabolic capacity implies a waste of energy;

usually it is replaced, when the cold exposure continues, by more economical modifications such as growth of fur.

Cold adaptation causes a marked increase in weight of brown adipose tissue (cf. p. 156) and a development of metabolically active multilocular cells (Heldmaier, 1974b). However, cells of brown adipose tissue from short daily cold exposed mice contained a small amount of cytoplasm and a large amount of fat concentrated in a single vacuole, in contrast to the multilocular cells seen during continuous cold exposure (Heldmaier, 1975).

In cold-adapted rats, the increase in turnover of free fatty acids following injection of noradrenaline (Moriya *et al.*, 1974; Maekubo *et al.*, 1977), as well as the sensitivity to the lipolytic activity of glucagon (Kuroshima *et al.*, 1975) are enhanced. It is assumed that, besides free fatty acids, ketone bodies are also used as energy source of non-shivering thermogenesis (Maekubo *et al.*, 1977).

Le Blanc and Villemaire (1970) have produced evidence for the significance of the thyroid in the development of brown fat of the rat by daily administration of thyroxine for 5 weeks; the growth of brown adipose tissue was stimulated still more by simultaneous administrations of noradrenaline. The weight of these tissues attained by use of the latter treatment was quite close to the weight found in the cold-adapted rat. After this treatment the animals were able to react to stimulation of the sympathetic system or to the injection of NA with an acute increase in heat formation that was even greater than in cold-adapted animals. It has been shown that local cooling of the preoptic hypothalamus stimulates thyroid activity (cf. p. 91). This mechanism might be involved in the formation of brown fat during prolonged cold exposure of small animals. By contrast, no answer has been found to the question of what the effects of the cold-induced increase in thyroxine secretion rate are in larger species, such as man, which do not form brown fat and are incapable of non-shivering thermogenesis.

15.6.2 PERIPHERAL BLOOD FLOW

Exposure to cold produces a less intense restriction of blood flow through the extremities of cold-adapted animals than through those of warm-adapted ones (Héroux and St Pierre, 1957; Harada and Kanno, 1975). These changes are, in part, due to an increase in the number of capillaries in the skin. In pigs raised at 25 °C ambient temperature, tail blood flow was considerably higher than in in animals raised at

35 °C, but the vascular response to various local hypothalamic temperatures was similar in both groups (Ingram, 1977).

When the human hand is immersed in ice water, after a period of dropping, the temperature begins to rise. This cold-induced vasodilatation (Lewis' reaction; cf. p. 164) has been shown to occur sooner and at higher temperatures in people whose hands had been repeatedly exposed to cold, such as various ethnic groups, fishermen, fish filleters, or subjects in laboratory experiments (for references see Hensel et al., 1973). It is not clear as to whether peripheral or central mechanisms are involved in these adaptive changes; it has also been claimed that physical fitness plays a major role, but in this respect the results are rather contradictory (Eagan, 1963; Livingstone, 1976).

Peripheral blood flow in humans (hand, forearm) increases in relation to body temperature by 10 to 30% in the course of heat adaptation (Fox et al., 1963) or a combination of exercise and heat adaptation (Roberts et al., 1977). This facilitates heat transport to the body surface. Adaptive changes in peripheral blood flow are associated with modifictions in central circulation, such as a relative decrease in heart rate and an increase in plasma volume (for references see Hensel et al., 1973; Senay, 1975; Senay et al., 1976; Wyndham et al., 1976; Bonner et al., 1976).

15.6.3 SWEAT SECRETION

The most fundamental modification during long-term heat adaptation of humans is to be seen in changes in the secretion of sweat and in the salt and water metabolism. Thus, if men are exposed to a continuous heat load or undergo repeated exposures, the sweat secretion rate, as measured under a standard heat stress, rises from day to day and may finally amount to about three times the initial value. This increase is partially due to a shift in sweating threshold to lower body temperatures and partially to a higher capacity of the sweat glands. The NaCl content of the sweat falls drastically in the course of heat adaptation and in extreme cases reaches 0.03%, i.e. ca. 1/25 of the NaCl concentration of the plasma (for references see Hensel et al., 1973).

There is evidence for the view that the increased sweating capacity is achieved mainly by an enhanced output of the sweat glands in response to the same nerve stimulus, although some contribution of the hypothalamus is not ruled out (Collins and Weiner, 1968). When the body temperature was repeatedly elevated by infrared heating but sweating was suppressed by bathing the lower part of the body in cold water, no adaptive increase in sweating capacity was seen in this region. On the other hand, a long-term increase in local sweat gland activity could be achieved by intradermal injection of acetyl-β-methylcholine on successive days without raising the body temperature. Thus the adaptive increase in sweating capacity seems to be due to a local training effect of secretory activity rather than to some changes in the CNS. It has further been shown that long-term heat adaptation is accompanied by a shift in regional distribution of sweating in such a way that the rate of secretion increases to a greater extent in the region of the extremities (Höfler, 1968) but we do not know whether these changes are central or peripheral in nature.

15.7 Changes in Thermal Comfort and Behaviour

15.7.1 COMFORT IN HUMANS

From the experience of everyday life it is presumed that man can adapt himself to prefer warmer or cooler surroundings. Thus people are said to prefer higher ambient temperatures in summer or in hot climates than they do in winter or in cold climates. However, physiological investigations in this field have led to rather controversial results.

According to Fanger and his co-workers (Fanger, 1970, 1973a,b,c,d; Fanger *et al.*, 1974a,b, 1977) the comfort conditions preferred by humans remain surprisingly stable under various long-term thermal stressors. Danes and Americans, people from the tropics, Danes working in the cold meat-packing industry, and winter swimmers showed practically the same comfort conditions as defined by preferred ambient temperatures (Table 15.II). Likewise, there were only minor or insignificant differences between men and women, day and night, winter and summer.

On the other hand, there is evidence from studies around the world that peoples' judgment of neutral temperature is displaced in the same direction as their thermal experience. For example, natives of certain hot-wet regions have been found to prefer warmer ambient conditions in summer seasons (Gonzalez, 1979). There is also some evidence for seasonal variations in the response to heat stress, in that the sweating reaction in summer was characterized by a relatively smaller salt loss despite a greater water loss and by corresponding changes in body fluids (Hori *et al.*, 1974; Morimoto *et al.*, 1974). Furthermore, it has been demonstrated (Brück *et al.*, 1976) that repeated cold exposure of human subjects leads to a marked decrease in cold discomfort and in general cold sensation at low ambient temperatures (cf. Fig. 15.5).

How can these apparent contradictions be explained? To a certain extent the differences seem to be a matter of the criteria for assessing thermal comfort. (1) In the experiments described by Fanger, the subjects chose conditions of maximal thermal comfort. In the experiments by Brück, Baum and Schwennicke the subjects estimated the

TABLE 15.II

Comfort conditions for different national-geographic groups and for groups of people regularly exposed to extreme cold or heat. Sedentary activity, clothing: 0.6, relative velocity $<0.1 \, \text{ms}^{-1}$, relative humidity: 50%, mean radiant temperature = air temperature

Group	Preferred ambient temp. (°C)	Mean skin temp. at comfort (°C)	Evaporative weight loss during comfort $(\text{g m}^{-2} \text{h}^{-1})$	Number of subjects
Americans	25.6			720
Danes	25.7			256
Danes	25.4	33.5	19.2	32
People from the tropics	26.2	33.5	17.1	16
Danes working in the cold meat-packing industry	24.7	33.6	17.1	16
Danish winter swimmers	25.0	33.3	16.6	16
Comfort equation	25.6			

From Fanger (1973b).

degree of discomfort at given ambient temperatures below the neutral value. If only the gain but not the threshold of the comfort function would change during long-term adaptation, the zone of maximal comfort may remain constant, whereas thermal discomfort outside the neutral zone may be diminished. The curves shown in Fig. 15.5 are compatible with this assumption. This would imply a certain dissociation between temperature regulation and thermal comfort, dissociation that has been shown to be possible in animals and humans (cf. Chapter 12 and 13). (2) In a multiple-loop control system, adaptive modifications may include a shift in the relation between central and peripheral thermal drives. Thus the set point for thermal comfort, as expressed by weighted mean body temperature, may deviate from the initial level in spite of a constant preferred skin or ambient temperature.

The present results on the adaptability of thermal comfort and their consequences for energy conservation can be summed up as follows: Adaptability of people to prefer low ambient temperatures is extremely limited. Therefore the only realistic way to save heating energy is to improve thermal insulation. Under hot climatic conditions the situation may be somewhat different. A certain adaptive shift of the preferred or accepted ambient temperature to higher levels seems possible, thereby saving appreciable amounts of energy for air-conditioning (Gonzalez, 1979).

15.7.2 BEHAVIOUR IN ANIMALS

In rats (Carlton and Marks, 1958), pigs and haired mice (Baldwin and Ingram, 1968c) exposed to cold for up to 2 weeks, an increase in the rate of work for heat was seen, increase which does not seem to be due to the fall in body weight induced by the cold exposure (Baldwin and Ingram, 1968c). This has been confirmed in pigs that were raised from 14 days of age at ambient temperatures of 10 and 35 °C, respectively (Heath and Ingram, 1978). The animals raised at 10 °C demanded more heat in cold environments than did the animals raised at 35 °C. On the other hand, rats sheared after 1 month of continuous cold exposure (Laties and Weiss, 1960) and hairless mice intermittently exposed to cold for a month (Revusky, 1966) showed a decrease in the rate of work for heat. These experiments seem to suggest that prolonged cold exposure, due to changes either in the thermal input or in the characteristics of the thermoregulatory system, increases the need for thermal behaviour and that a decreased need for thermal behaviour appears only after the animals become adapted to cold. This interpretation seems to be supported by Revusky's data (1966) which show a big rate of work for heat during the second week of intermittent cold exposure and a progressive decline in the rate of thermal behaviour in consecutive days.

Rats were trained, in a cold environment, to press a lever to obtain heat. Local hypothalamic heating suppressed this response (Carlisle, 1966). When the experiment was repeated over a period of several days, the effect of hypothalamic heating on behaviour diminished, that is, working for heat continued at a somewhat slower rate during hypothalamic heating. According to Carlisle, it cannot be decided whether this may be attributed to tissue damage as a result of heating, or to experience in this situation.

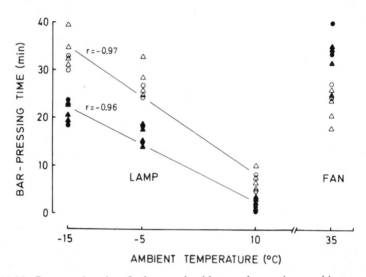

Fig. 15.12. Bar-pressing time for heat and cold reward at various ambient temperatures in control (solid symbols) and experimental rats (open symbols), with preoptic (triangles) and spinal cord (circles) thermodes that had been intermittently cooled for several days before the experiment started. (From Cormarèche-Leydier et al., 1977.)

The thermoregulatory behaviour of rats was studied during exposure to cold and warm ambient temperatures after either the preoptic area or the cervical spinal cord had been intermittently cooled for an average of 130 h (Cormarèche-Leydier et al., 1977). The precooled animals worked more for heat in cold environments and less for cool air in a warm environment than the control animals (Fig. 15.12). This behaviour, probably due to a decreased ability of the precooled animals to retain heat, suggests that the precooled animals were not fully adapted to cold.

16
Ontogenesis of Temperature Regulation

16.1 Temperature Regulation in Infants

16.1.1 BODY TEMPERATURE

Many homeothermic neonates have approximately the same average level of body temperature as found in adults (for references see Hensel *et al.*, 1973; Brück, 1978b; Blix and Steen, 1979). This holds also for the human neonate, with the exception of the first few hours of life, during which thermal responses are depressed and internal body temperature may be slightly below the adult level. The somewhat sluggish reaction is probably due to birth stress; this is inferred by the fact that metabolic depression was more pronounced in infants who had undergone severe delivery (Brück, 1961).

The diurnal temperature fluctuation of body temperature is absent in the human neonate and develops during the first weeks of life (Hellbrügge, 1960); in children it amounts to 1.7 °C, which is greater than in adults. Stability of the average level of internal body temperature in the growing organism implies considerable changes in the thermoregulatory system during ontogenesis.

16.1.2 BODY SIZE AND METABOLISM

The neonatal passive system is characterized by a large body surface/volume ratio and a decreased thickness of the body shell. A comparison of values for the human neonate (Hey *et al.*, 1970) and adult (Hardy, 1961) shows that overall insulation changes by a factor 1.8 when body mass increases from 3.5 to 70 kg. Tissue insulation increases by a factor 3.0, and the air insulation, due to the larger

241

curvature radii of trunk and extremities, by a factor 1.2. The surface/volume ratio decreases with growth; for the full-term neonate it is 2.7 times and for the 1-kg premature neonate four times larger than for adults (Brück, 1978b).

The heat production (H) per body mass (W) required for equilibrium heat flow at any given overall temperature difference would

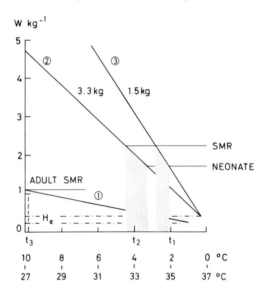

FIG. 16.1. Schematic representation of heat production per unit body mass required to maintain the temperature differences (ΔT) between body core (37 °C) and environment $(T_o,$ operative temperature) in the adult (1) and in a 3.3-kg (2) and a 1.5-kg (3) infant (with maximum vasoconstriction). With the actual SMR given, the temperature differences t_1, t_2, and t_3 can be maintained under conditions of thermal neutrality; conversely, the neonates require an environmental temperature of about 33 °C (2) or even 34 to 35 °C (3) to maintain a deep-body temperature of 37 °C, while in the adult this figure is much lower, 27 °C (t_3). H_e, Minimum evaporative heat loss. (From Brück, 1978b.)

thus have to be nearly five times as large in the neonate than in the adult. The equation

$$H/W = k \cdot W^{0.5} \tag{24}$$

approximately predicts H/W in relation to body mass at a given temperature difference between body core and environment. In man this equation holds from the second year of life on, corresponding to a body mass of over 10 kg. In the first year, in particular in the first week of life, the standard metabolic rate (SMR) per unit of body mass is only 1.5 to 2 times larger than in the adult (Brück, 1978c). This means that the overall temperature difference (ΔT) which can be held in the neonate under the condition of SMR is only less than half that of the adult, as shown by Fig. 16.1. At an ambient temperature below 33 °C, deep body temperature would drop unless heat production is increased above SMR. This actually occurs at this high ambient temperature level, that is, the lower limit of the neutral temperature range is shifted to a higher level of ambient temperature.

16.1.3 DEVIATION IN THERMOREGULATORY THRESHOLDS

From a theoretical point of view, the stability of body temperature in the neonate requires an input/output relationship of the thermoregulatory system which differs from that of adults. If this were not the case, the higher cold-induced metabolic rate in the neonate would require a higher load error (cf. p. 16), and thus the body temperature

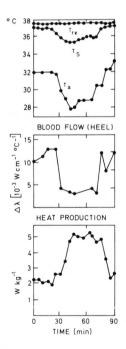

FIG. 16.2. Response of a 7-day-old human neonate to external cooling. T_{re}, rectal temperature; $\bar{T}_s$, mean skin temperature; T_a, ambient temperature. (From Brück, 1961.)

at a given ambient below 33 °C would be lower in the neonate than in the adult, which obviously is not the case. In fact, cold defence reactions in the neonate are elicited at higher mean skin temperatures than in the adult.

In human neonates only a few days old, heat production more than doubles and vasoconstriction occurs while skin temperature falls to only 35 °C; the corresponding ambient temperature being 28 °C. Deep body temperature remains constant (Fig. 16.2).

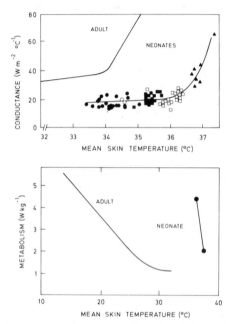

Fig. 16.3 Average thermogenetic responses of human adults and neonates in relation to mean skin temperature. Conductance in adults: Hardy *et al.* (1970); in neonates: Ryser and Jéquier (1972); metabolism in neonates: Brück (1961). (From Brück, 1978c.)

Extra heat production and maximum vasoconstriction are elicited and maintained in the newborn at much higher skin and mean body temperature than in the adult (Hey and Katz, 1970; Hey *et al.*, 1970; Brück, 1978a,b,c). In addition, the increment of oxygen uptake in relation to average skin temperature is larger in the neonate than in the adult (Fig. 16.3).

This shift in threshold for metabolic and vasomotor responses to higher mean body temperature can be considered as an adjustment to the small body size. According to Brück (1978a), the threshold deviation in neonates may be explained in two ways: (i) due to the increased body surface/body mass quotient the number of cold receptors per unit of body mass is increased, assuming that the density of the distribution and the characteristics of cold receptors are the same in adults and neonates; or (ii) the central processing of thermal input signals is different in the neonate from that in the adult. In any case, the increased cutaneous sensitivity is prerequisite for enabling the newborn infant to maintain its core temperature at the same level and with the same accuracy as does an adult.

In young premature infants (below 1500 g), the threshold for metabolic increment and vasoconstriction is shifted to deep body temperature of around 35 °C, while no downward shift of the sweating threshold has been found (Brück, 1978a). Thus, with respect to the position of thresholds for actuation of thermoregulatory effectors, the small human neonate resembles the intermittently cold-exposed guinea pig in which a widening of the interthreshold zone has been demonstrated (cf. p. 228).

16.1.4 CONTROL OF EFFECTOR SYSTEMS

Figure 16.4 shows a schematic diagram of neonatal temperature regulation, based on studies in the guinea pig (Brück and Wünnenberg, 1970; Brück, 1978b). In particular, the interlocked control of shivering and non-shivering thermogenesis as well as the special feedback mechanism through preferential vascular heat transfer from the interscapular brown adipose tissue to the spinal cord receptor area are shown. Increasing heat production of the brown adipose tissue will lead to a temperature rise of the spinal cord and thus to an activation of hypothalamic neurons that inhibit shivering. This diagram may, in part, also apply to the human neonate. Shivering in neonates who exhibit non-shivering thermogenesis does not occur before non-shivering thermogenesis has been induced to its full extent.

Non-shivering thermogenesis plays an important role as a source of cold-induced heat production in the neonates of many species of mammals including the human infant (for references see Brück, 1970a, 1978b). As an exception, the pig and miniature pig may be mentioned. Newborn piglets have neither brown adipose tissue, nor do they develop non-shivering thermogenesis (Mount, 1968; Brück, 1978b).

The distribution of brown adipose tissue in the newborn human infant (Merklin, 1974) is shown in Fig. 16.5. The interscapular

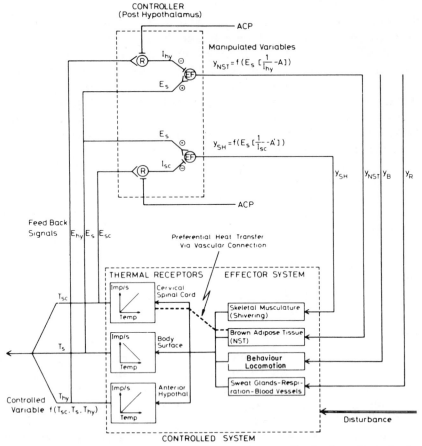

FIG. 16.4 Neonatal temperature regulation, based on studies in the guinea pig. In particular, the interlocked control of shivering and non-shivering thermogenesis as well as the special feedback mechanism through preferential vascular heat transfer from the interscapular adipose tissue to the spinal cord receptor area are shown. The central integrative mechanisms for heat dissipation and behavioural control (y_R and y_B) have not yet been worked out in the neonate. + and —⊂, Facilitation; − and ⊣, inhibition; T, temperature; imp, impulse frequency in nerves; s, body surface; hy, hypothalamus; sc, spinal cord; E, facilitatory or excitatory impulses; I, inhibitory impulses; R, interneurons whose spontaneous activity is thought to influence the reference signal; EF, effector neuron; ACP, ascending catecholaminergic pathway influencing level of reference signals. A and A' are constants less than 1. (From Brück, 1978a.)

adipose tissue is rather diffusely distributed, whereas a compact fat pad is found in various rodents (cf. p. 156).

The following results show the existence of non-shivering thermogenesis in the human neonate: (1) The newborn infant responds to intravenous administration of noradrenaline (NA) with an increase in oxygen consumption, but without an appreciable increase in physical activity (Karlberg *et al.*, 1965). (2) Free fatty acids and glycerol are liberated on cold exposure of the neonate and also after NA infusion (Schiff *et al.*, 1966; Hull and Hardman, 1970). (3) Renal NA excretion increases simultaneously with the increase in O_2 uptake in cold-exposed newborn infants, while adrenaline excretion remains small (Stern *et al.*, 1965). This indicates that cold-defence reactions are mediated by sympathetic nerve terminals. (4) Skin temperature in the dorsal interscapular area is relatively increased in the human infant when a cold-induced metabolic increase is provoked (Silverman *et al.*, 1964).

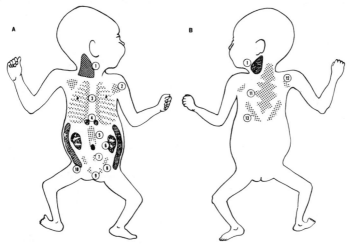

FIG. 16.5. Distribution of brown adipose tissue in the human neonate. (A) Frontal and (B) dorsal views. (From Merklin, 1974.)

When rats from birth up to the 14th day were intermittently exposed to ambient temperatures of 4 °C, the adaptive increase in non-shivering thermogenesis lasted for about 18 weeks, whereas the same cold exposure in adult rats was followed by a readaptation period of only 4 weeks (Doi and Kuroshima, 1979). Thus the lasting effect of cold adaptation is much longer in newborn animals than in adults. For the development of non-shivering thermogenesis in the infant rat, the thyroid gland seems to be essential. Thus suppression of thyroxine formation by administration of propylthiouracil was followed by a marked decrease in the non-shivering thermogenetic response to cold or NA (Steele and Wekstein, 1972, 1973).

As a consequence of their physical properties, such as a large body surface/body mass ratio and a high thermal conductivity of the body shell, human neonates are not able to maintain a normal body temperature at low ambient temperatures. In warm environments, the maximum evaporative heat loss of full-term neonates was found to be only in the range of SMR (Hey and Katz, 1969; Hey, 1972; Sulyok, et al., 1974), whereas in the adult it amounts to five times SMR. Thus the human neonate must be considered to be markedly less resistant to high and low temperatures than the adult.

No sweating to thermal stimuli could be detected in premature infants of less than 210 days post-conceptional age (Hey and Katz, 1969). In such infants, the sweat glands did not even respond to intradermal injection of acetylcholine. It is assumed that the failing sweating response is mainly due to an immaturity of the sweat glands (Foster et al., 1969). In contrast, even in the very small premature, we find metabolic and vasomotor thermoregulatory responses (Brück, 1978a,b).

Temperature regulation is frequently referred to as being "not fully developed" at the time of birth. We should be very cautious, however, in asserting the immaturity of the control system, even though the neonate shows more fluctuations in body temperature than the adult. As yet, there is no indication for a general immaturity of the nervous or hormonal systems that are involved in the transfer and integration of thermal drives. Thus instability of the body temperature seems to be due mainly to the discrepancy between the efficiency of effector systems and body size (Brück, 1978a,b). This statement holds for normal temperature regulation; it does not imply that there are no differences whatsoever in the central thermoregulatory processes between neonates and adults. As discussed in the following, such differences have been found for the response to pyrogens.

16.1.5 RESPONSE TO PYROGENS

There are numerous reports in the literature of newborn and premature infants suffering severe infections, for example meningitis, enteritis and septicaemia, without the expected accompaniment of a marked rise in body temperature (Epstein et al., 1951; Smith et al., 1956; Craig, 1963).

Guinea pigs under 8 days of age generally are unable to develop fever when moderate doses of endotoxin or endogenous pyrogen are administered (Blatteis, 1975, 1976a, 1977; Blatteis and Smith, 1979). In guinea pigs during the first two to three weeks of life, non-shivering thermogenesis plays a major role in the fever response, as shown by the administration of propanolol. This β-adrenergic blocking agent suppresses the increased non-shivering thermogenesis caused by the

administration of endotoxin (Blatteis, 1976a,b). A diminished autonomic response to pyrogens in the early postnatal stages has also been
found in rabbits (Satinoff *et al.*, 1976a,b; Lipton and Ticknor, 1979)
and lambs (Pittman *et al.*, 1974; Cooper *et al.*, 1975; Pittman *et al.*,
1975, 1977; Kasting *et al.*, 1979).

The question arises as to whether newborn animals are not able to
produce endogenous pyrogen (EnP) or whether some pyrogen-
sensitive structures are not developed (cf. p. 211). It could be

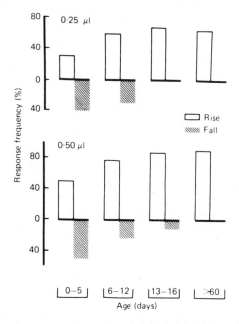

FIG. 16.6. Percent frequency distribution of guinea pigs 0 to 5, 6 to 12, 13 to 16, and
more than (>) 60 days of age reacting with either a rise (open bars) or a fall (hatched
bars) in body temperature ($\Delta T > 0.5$ °C) to 0.25 (upper bars) and 0.50 (lower bars)
μl intrapreoptic leucocytic pyrogen. (From Blatteis and Smith, 1979.)

demonstrated that newborn guinea pigs develop fever after high doses
of endotoxin and are thus able to elaborate EnP (Blatteis, 1977).
Similar observations have been made in newborn lambs (Pittman
et al., 1975, 1977); in other cases, the lambs did not respond to
moderate doses of endotoxin and became hypothermic with high
doses (Kasting *et al.*, 1979).

When EnP was microinjected into the preoptic area of newborn
guinea pigs, the animals responded either with a small rise or with a

fall in body temperature (Blatteis and Smith, 1979). As shown in Fig. 16.5, the hypothermic response ceased and the fever response increased with growing age of the animals. We may thus conclude that febrile responsiveness may depend on the stage of development of, presumably, the pyrogen-receptive mechanism. The results further imply that the preoptic sites where EnP acts and thermoafferents are integrated may not be the same, since the thermoregulatory capability in guinea pigs is fully competent from birth.

In newborn lambs, intraventricular injection of prostaglandin E_1 (Pittman et al., 1975), or intrahypothalamic injection of prostaglandin E_1 and E_2 (Pittman et al., 1977) did not cause fever, although the animals were able to develop fever after i.v. injection of bacterial pyrogen. This is further support for the view that prostaglandin is not a mediator of pyrogen fever (cf. p. 214).

Newborn rabbits did not develop a fever when injected intraperitoneally with a pyrogen and maintained at an ambient temperature of 32 °C (Satinoff et al., 1976a,b). When placed in a thermal gradient, however, animals injected with pyrogens selected gradient positions that represented significantly higher temperatures (40.4 °C) than controls with saline (36.4 °C). Allowing the rabbit pups to remain at their selected positions for 5 min caused a significant rise in the rectal temperatures of pyrogen-injected animals but not in that of controls. Thus, newborn rabbits will develop a fever by behavioural means after administration of exogenous pyrogens. This observation demonstrates once more a dissociation between autonomic temperature regulation and thermoregulatory behaviour.

16.2 Temperature Regulation and Comfort in Men and Women

Comparative studies of men and women during heat exposure have shown that men begin sweating at lower internal temperatures (Fox et al., 1969; Bittel and Henane, 1975). Peripheral circulation at 38 °C tympanic temperature was found to be higher in men than in women (Fox et al., 1969), and the threshold for forearm vasodilatation was shifted to lower oesophageal temperatures in men (Roberts et al., 1977). Men and women show no difference in rectal temperature during cold exposure, but women are able to reduce core-to-skin thermal conductance more than men, and so have cooler skins (Hardy and Du Bois, 1940; Wyndham et al., 1964). There was practically no difference in metabolism between men and women in ambient temperatures of either 27 or 5 °C (Wyndham et al., 1964).

In a multiple-loop control system, the measurement of a single thermoregulatory parameter, or the assessment of the relationship between a single thermal drive (e.g. rectal temperature) and a control action may not be sufficient to reveal possible changes in the characteristics of the thermoregulatory system. Instead, it is necessary to account for the combined thermal drive from various sites of the body. A satisfactory approximation of this integrated or "central" drive is the weighted mean body temperature (cf. p. 11).

Thermoregulatory responses as function of the "central drive" were investigated in men and women by Cunningham *et al.* (1978). The central drive (T'_c) was defined as

$$T'_c = T_{ty} + 0.1\,(\bar{T}_s - 33\,°\mathrm{C}) \tag{25}$$

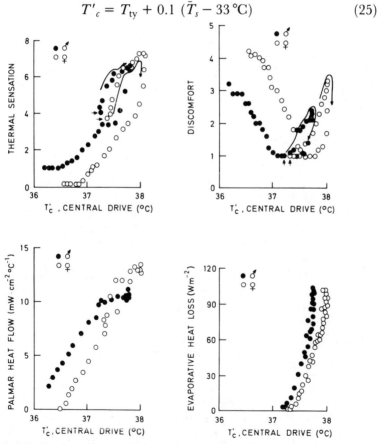

Fig. 16.7. Magnitude estimates of thermal sensation and discomfort, palmar heat conductance and evaporative heat loss as a function of central thermal drive in men and women. Small horizontal and vertical arrows indicate initial values. Successive data points of thermal sensation and discomfort describe an open loop clockwise with rising and falling temperature. (From Cunningham *et al.*, 1978.)

where T_{ty} is tympanic temperature and $\bar{T}_s$ mean skin temperature. This expression is based on the relative weightings of core and mean skin temperature that have best accounted for previous measurements of sweating and peripheral blood flow (Nadel et al., 1971b; Wenger et al., 1975b). Starting from neutral ambient temperatures, the temperature was raised to 48 °C and then lowered to 16 °C. Figure 16.7 shows some of the results. Evaporative heat loss and palmer heat conductance have lower thresholds of the central drive in men than in women. Likewise, the thresholds of thermal discomfort and of estimated magnitude of thermal sensation are found at lower levels of the central drive in men. Some of the functions show a clockwise hysteresis indicating a dynamic component when the temperature is changed with time in an upward or downward direction. Women's thermoregulatory responses were similar to the men's, but were shifted by about +0.3 °C for responses to heat and +0.6 °C for responses to cold. Women thus respond less to elevated mean body temperature and more to lowered temperatures than men.

16.3 Thermal Responsiveness in the Elderly

Research into the biology of ageing has shown there to be a decline in the efficiency of theremoregulatory functions with older age. Elderly people are less able to compensate for heat loss to cold environments than are younger subjects. Younger men responded with a higher metabolic increase and a stronger cutaneous vasoconstriction to a standard cold stress, thereby keeping their internal body temperature at higher levels than did older men (Wagner et al., 1974). This difference may be due to a lower threshold temperature as well as a lower capacity of heat-conservation responses in the elderly. In cold environments, people of high age are thus in danger of becoming hypothermic (Fox et al., 1973; Collins et al., 1977). A survey in Great Britain has shown that hypothermia of less than 35 °C body core temperature occurs in 0.58% of the population over 65 years of age (Collins, 1979).

The response to heat declines as well with old age. Older boys (15 to 16 years) and young men (25 to 30 years) showed a higher sensitivity of the sweating response and a higher secretory capacity of the sweating mechanism than older men (46 to 67 years). However, preadolescent boys (11 to 14 years) had a similar sweating response as seen in the elderly (Wagner et al., 1972).

It is generally assumed that older people prefer a somewhat higher ambient temperature than younger ones. The American Society of

Heating, Refrigerating and Air-Conditioning Engineers (ASHRAE, 1966) recommends room temperatures about 0.5 °C higher for people over 40. However, experiments under standard conditions have revealed comfort temperatures to be practically identical for young and old people (Rohles, 1969; Fanger, 1977; Langkilde, 1979). Nevertheless, the suggestion of a somewhat higher ambient comfort temperature for elderly people seems justified because this group, on the average, is physically less active than younger people (Langkilde, 1979).

Among elderly females living at mean daytime ambient temperatures of 18 °C, mainly because they could not afford more heating, about half of the subjects did not complain about cold discomfort (Watts, 1972). This may be explained by a certain degree of adaptive cold tolerance due to a high incidence of hypothermia rather than by a lower temperature of maximal thermal comfort due to high age.

Only very few studies of thermal perception in the elderly are as yet available. According to Kenshalo (1977, 1979a,b) there is no systematic difference in the thresholds of warm and cold sensation between young persons and subjects over 70 years of age. In a recent study, no significant difference between subjects of 25 and over 70 was found when warm and cold thresholds at the palm of the hand were estimated (Blatteis, Hensel and Issing, unpublished), although there seemed to be a slight tendency towards an increased cold sensitivity and a decreased warm sensitivity in the elderly.

References

ADAIR, E. R. (1969). Hypothalamic control of behavioral temperature regulation in the squirrel monkey. *Physiologist* **12,** 155.

ADAIR, E. R. (1970). Control of thermoregulatory behavior by brief displacements of hypothalamic temperature. *Psychon. Sci.* **20,** 11–13.

ADAIR, E. R. (1971). Evaluation of some controller inputs to behavioral temperature regulation. *Int. J. Biometeorol.* **15,** 121–128.

ADAIR, E. R. (1974). Hypothalamic control of thermoregulatory behavior. *In* "Recent Studies of Hypothalamic Function", pp. 341–358. Karger, Basel.

ADAIR, E. R. (1976a). Autonomic thermoregulation in squirrel monkey when behavioral regulation is limited. *J. appl. Physiol.* **40,** 694–700.

ADAIR, E. R. (1976b). Behavioral temperature regulation in the squirrel monkey: unilateral versus bilateral preoptic thermal stimulation. *Israel J. med. Sci.* **12,** 1033–1035.

ADAIR, E. R. (1977a). Studies on the behavioral regulation of preoptic temperature. *In* "Drugs, Biogenic Amines and Body Temperature", pp. 84–86. Karger, Basel.

ADAIR, E. R. (1977b). Skin, preoptic, and core temperatures influence behavioral thermoregulation. *J. appl. Physiol.* **42,** 559–564.

ADAIR, E. R. and RAWSON, R. O. (1973). Step-wise changes in thermoregulatory responses to slowly changing thermal stimuli. *Pflügers Arch.* **339,** 241–250.

ADAIR, E. R. and RAWSON, R. O. (1974). Autonomic and behavioral temperature regulation unilateral vs bilateral preoptic thermal stimulation. *Pflügers Arch.* **352,** 91–103.

ADAIR, E. R. and WRIGHT, B. A. (1976). Behavioral thermoregulation in the squirrel monkey when response effort is varied. *J. comp. physiol. Psychol.* **90,** 179–184.

ADAIR, E. R., CASBY, J. U. and STOLWIJK, J. A. J. (1970). Behavioral temperature regulation in the squirrel monkey: changes induced by shifts in hypothalamic temperature. *J. comp. physiol. Psychol.* **72,** 17–27.

ADOLPH, E. F. (1964). Perspectives of adaptation: some general properties. *In* "Handbook of Physiology", Sect. IV, pp. 27–35. Amer. Physiol. Soc., Washington, D.C.

254

ALBE-FESSARD, D. and BESSOU, J. M. (1973). Convergent thalamic and cortical projections—the non-specific system. In "Handbook of Sensory Physiology" (A. Iggo, Ed.), Vol. 2, pp. 489–560. Springer, Berlin, Heidelberg and New York.

AL-HACHIM, G. M. and FRENS, J. (1975). Analysis of shivering in nonperipheral cooling during pyrogen fever. Int. J. Biometeorol. 19, 53–55.

ALLEN, J. A. and RODDIE, I. C. (1972). The role of circulating catecholamines in sweat production in man. J. Physiol. (Lond.) 227, 801–814.

ALRUTZ, S. (1897). Studien auf dem Gebiet der Temperatursinne. I. Skand. Arch. Physiol. 7, 321–340.

ALRUTZ, S. (1898). On the temperature senses: II. The sensation "hot". Mind 7, ser. 2, 140–144.

ALRUTZ, S. (1900). Studien auf dem Gebiet der Temperatursinne. II. Skand. Arch. Physiol. 10, 340–352.

AMERICAN SOCIETY OF HEATING, REFRIGERATING AND AIR-CONDITIONING ENGINEERS (1966). Thermal comfort conditions. ASHRAE Standard, 55–66.

ANDERSSON, B. (1970). Central nervous and hormonal interaction in temperature regulation of the goat. In "Physiological and Behavioral Temperature Regulation" (J. D. Hardy, A. P. Gagge and J. A. J. Stolwijk, Eds), pp. 634–647. Thomas, Springfield, Ill.

ANDERSSON, B., EKMAN, L., GALE, C. C. and SUNDSTEN, J. W. (1962). Blocking of the thyroid response to cold by local warming of the preoptic region. Acta physiol. scand. 56, 94–96.

ANDERSSON, B., GALE, C. C., HÖKFELT, B. and OHGA, A. (1963). Relation of preoptic temperature to the function of the sympathico-adrenomedullary system and the adrenal cortex. Acta physiol. scand. 61, 182–191.

ANDERSSON, B., BROOK, A. H. and EKMAN, L. (1965a). Further studies of the thyroidal response to local cooling of the "heat loss center". Acta physiol. scand. 63, 186–192.

ANDERSSON, B., GALE, C. C., HÖKFELT, B. and LARSSON, B. (1965b). Acute and chronic effects of preoptic lesions. Acta physiol. scand. 65, 45–60.

APPLEBAUM, A. E., BEALL, J. E., FOREMAN, R. D. and WILLIS, W. D. (1975). Organization and receptive fields of primate spinothalamic tract neurons. J. Neurophysiol. 38, 572–586.

ASCHOFF, J. (1970). Circadian rhythm of activity and of body temperature. In "Physiological and Behavioral Temperature Regulation" (J. D. Hardy, A. P. Gagge and J. A. J. Stolwijk, Eds), pp. 905–919. Thomas, Springfield, Ill.

ASCHOFF, J. (1979). Circadian rhythms: general features and endocrinological aspects. In "Endocrine Rhythms" (D. T. Krieger, Ed.), pp. 1–61. Raven Press, New York.

ASHBY, W. R. (1966). Mathematical models and computer analysis of the function of the central nervous system. Ann. Rev. Physiol. 28, 89–106.

BACON, M. and BLIGH, J. (1976). Interaction between the effects of spinal heating and cooling and of injections into a lateral cerebral ventricle of

256   THERMORECEPTION AND TEMPERATURE REGULATION

noradrenaline, 5-hydroxytryptamine and carbachol on thermoregulation in sheep. *J. Physiol. (Lond.)* **254**, 213–227.

BADE, H., BRAUN, H. A., HENSEL, H. and SCHÄFER, K. (1978). Discharge pattern of cold fibres related to hypothetical receptor mechanisms. *J. Physiol. (Lond.)* **284**, 83 P.

BADE, H., BRAUN, H. A. and HENSEL, H. (1979). Parameters of the static burst discharge of lingual cold receptors in the cat. *Pflügers Arch.* **382**, 1–5.

BAKER, M. A. and HAYWARD, J. N. (1967). Autonomic basis for the rise in brain temperature during paradoxical sleep. *Science* **157**, 1586–1588.

BAKER, M. A., STOCKING, R. A. and MEEHAN, J. P. (1972). Thermal relationship between tympanic membrane and hypothalamus in conscious cat and monkey. *J. appl. Physiol.* **32**, 739–742.

BALDWIN, B. A. (1975). The effects of intra-ruminal loading with cold water on thermoregulatory behaviour in sheep. *J. Physiol. (Lond.)* **249**, 139–152.

BALDWIN, B. A. and INGRAM, D. L. (1967). The effect of heating and cooling the hypothalamus on behavioural thermoregulation in the pig. *J. Physiol. (Lond.)* **191**, 375–392.

BALDWIN, B. A. and INGRAM, D. L. (1968a). The influence of hypothalamic temperature and ambient temperature on thermoregulatory mechanisms in the pig. *J. Physiol. (Lond.)* **198**, 517–529.

BALDWIN, B. A. and INGRAM, D. L. (1968b). Factors influencing behavioural thermoregulation in the pig. *Physiol. Behav.* **3**, 409–415.

BALDWIN, B. A. and INGRAM, D. L. (1968c). The effects of food intake and acclimatization to temperature on behavioural thermoregulation in pigs and mice. *Physiol. Behav.* **3**, 395–400.

BALDWIN, B. A. and YATES, J. O. (1977). The effects of hypothalamic temperature variation and intracarotid cooling on behavioural thermo-regulation in sheep. *J. Physiol. (Lond.)* **265**, 705–720.

BALDWIN, B. A., INGRAM, D. L. and LEBLANC, J. (1969). The effects of environmental temperature and hypothalamic temperature on excretion of catecholamines in the urine of pigs. *Brain Res.* **16**, 511–515.

BALLANTYNE, E. R., HILL, R. K. and SPENCER, J. W. (1977). Probit analysis of thermal sensation assessments. *Int. J. Biometeorol.* **21**, 29–43.

BANET, M. (1979). Fever and survival in the rat. *Pflügers Arch.* **381**, 35–38.

BANET, M. and HENSEL, H. (1976a). The interaction between cutaneous and spinal thermal inputs in the control of oxygen consumption in the rat. *J. Physiol. (Lond.)* **260**, 461–473.

BANET, M. and HENSEL, H. (1976b). Nonshivering thermogenesis induced by repetitive hypothalamic cooling in the rat. *Amer. J. Physiol.* **230**, 522–526.

BANET, M. and HENSEL, H. (1976c). Nonshivering thermogenesis induced by repetitive cooling of spinal cord in the rat. *Amer. J. Physiol.* **230**, 720–723.

BANET, M., HENSEL, H. and LIEBERMANN, H. (1978a). The central control of shivering and non-shivering thermogenesis in the rat. *J. Physiol. (Lond.)* **283**, 569–584.

BANET, M., HENSEL, H. and POTHMANN, B. (1978b). Autonomic thermo-regulation after intermittent cooling of the spinal cord and cold exposure in the rat. *J. Physiol. (Lond.)* **275,** 439–447.

BANKS, W. P. (1973). Reaction time as a measure of summation of warmth. *Percept. Psychophys.* **13,** 321–327.

BARCROFT, H., BOCK, K. D., HENSEL, H. and KITCHIN, A. H. (1955). Die Muskeldurchblutung des Menschen bei indirekter Erwärmung und Abkühlung. *Pflügers Arch. ges. Physiol.* **261,** 199–210.

BARGMANN, W., HEHN, G. VON and LINDNER, E. (1968). Über die Zellen des braunen Fettgewebes und ihre Innervation. *Z. Zellforsch.* **85,** 601–613.

BARKER, J. L. and CARPENTER, D. O. (1970). Thermosensitivity of neurons in the sensorimotor cortex of the cat. *Science* **169,** 597–598.

BARKER, J. L. and CARPENTER, D. O. (1971). Neuronal thermosensitivity. *Science* **172,** 1361–1362.

BARNEY, C. C. and ELIZONDO, R. S. (1978). Effect of ambient temperature on development of prostaglandin E_1 hyperthermia in the rhesus monkey. *J. appl. Physiol.* **44,** 751–758.

BAUM, E., BRÜCK, K. and SCHWENNICKE, H. P. (1976). Adaptive modifica-tions in the thermoregulatory system of long-distance runners. *J. appl. Physiol.* **40,** 404–410.

BAZETT, H. C. and McGLONE, B. (1930). Experiments on the mechanism of stimulation of end-organs for cold. *Amer. J. Physiol.* **93,** 632.

BAZETT, H. C., McGLONE, B. and BROCKLEHURST, R. J. (1930). The temperatures in the tissues which accompany temperature sensations. *J. Physiol. (Lond.)* **69,** 88–112.

BECK, P. W. and HANDWERKER, H. O. (1974). Bradykinin and serotonin effects on various types of cutaneous nerve fibres. *Pflügers Arch.* **347,** 209–222.

BECK, P. W., HANDWERKER, H. O. and ZIMMERMANN, M. (1974). Nervous outflow from the cat's foot during noxious radiant heat stimulation. *Brain Res.* **67,** 373–386.

BECKMAN, A. L. and EISENMAN, J. S. (1970). Microelectrophoresis of biogenic amines on hypothalamic thermosensitive cells. *Science* **170,** 334–336.

BEDFORD, T. (1936). The warmth factor in comfort at work. *Rep. industr. Hlth. Res. (Lond.)* No. 76.

BEHLING, K., BLEICHERT, A., KITZING, J., NINOW, S., SCARPERI, M. and SCARPERI, S. (1971). An analog model of thermoregulation during rest and exercise. *Int. J. Biometeorol.* **15,** 212–218.

BEITEL, R. E. and DUBNER, R. (1976). Response of unmyelinated (C) polymodal nociceptors to thermal stimuli applied to monkey's face. *J. Neurophysiol.* **39,** 1160–1175.

BEITEL, R. E., DUBNER, R., HARRIS, R. and SUMINO, R. (1977). Role of thermoreceptive afferents in behavioral reaction times to warming temper-ature shifts applied to the monkey's face. *Brain Res.* **138,** 329–346.

BENZINGER, T. H. (1959). On physical heat regulation and the sense of temperature in man. *Proc. nat. Acad. Sci. (Wash.)* **45,** 645–659.

258   THERMORECEPTION AND TEMPERATURE REGULATION

BENZINGER, T. H. (1969). Heat regulation: Homeostasis of central temperature in man. *Physiol. Rev.* **49,** 671–759.

BENZINGER, T. H. (1970). Peripheral cold reception, sensory mechanism of behavioral and autonomic thermostasis. *In* "Physiological and Behavioral Temperature Regulation" (J. D. Hardy, A. P. Gagge and J. A. J. Stolwijk, Eds), pp. 831–855. Thomas, Springfield, Ill.

BENZINGER, T. H. (1979). The physiological basis for thermal comfort. *In* "Indoor Climate" (P. O. Fanger and O. Valbjørn, Eds), pp. 441–476. Danish Building Res. Inst., Copenhagen.

BENZINGER, T. H., KITZINGER, C. and PRATT, A. W. (1963). The human thermostat. *In* "Temperature, its Measurement and Control in Science and Industry" (J. D. Hardy, Ed.), Vol. 3, pp. 637–665. Reinhold, New York.

BERGLUND, L. G. (1979). Occupant acceptance of temperature drifts. *In* "Indoor Climate" (P. O. Fanger and O. Valbjørn, Eds), pp. 507–525. Danish Building Res. Inst., Copenhagen.

BESSOU, P. and PERL, E. R. (1969). Response of cutaneous sensory units with unmyelinated fibers to noxious stimuli. *J. Neurophysiol.* **32,** 1025–1043.

BESTE, R. (1977). Perzeption statischer thermischer Reize beim Menschen. Inaug.-Diss., Marburg.

BESTE, R. and HENSEL, H. (1977). Subjective estimation of static temperatures at the palm in humans. *Pflügers Arch.* **368,** R47.

BESTE, R. and HENSEL, H. (1978). Pleasantness of local static thermal stimuli during changing mean skin temperature. *Pflügers Arch.* **373,** R90.

BIRZIS, L. and HEMINGWAY, A. (1957). Efferent brain discharge during shivering. *J. Neurophysiol.* **20,** 156–166.

BITTEL, J. and HENANE, R. (1975). Comparison of thermal exchanges in men and women under neutral and hot conditions. *J. Physiol. (Lond.)* **250,** 475–489.

BJÖRKLUND, A. and NOBIN, A. (1973). Fluorescence histochemical and microspectrofluroscopic mapping of dopamine and noradrenaline cell groups in the rat diencephalon. *Brain Res.* **51,** 193–205.

BLATTEIS, C. M. (1960). Afferent initiation of shivering. *Amer. J. Physiol.* **199,** 697–700.

BLATTEIS, C. M. (1975). Postnatal development of pyrogenic sensitivity in guinea pigs. *J. appl. Physiol.* **39,** 251–257.

BLATTEIS, C. M. (1976a). Fever: exchange of shivering by non-shivering pyrogenesis in cold-acclimated guinea pigs. *J. appl. Physiol.* **40,** 29–34.

BLATTEIS, C. M. (1976b). Effect of propranolol on endotoxin-induced pyrogenesis in newborn and adult guinea pigs. *J. appl. Physiol.* **40,** 35–39.

BLATTEIS, C. M. (1977). Comparison of endotoxin and leukocytic pyrogen pyrogenicity in newborn guinea pigs. *J. appl. Physiol.* **42,** 355–361.

BLATTEIS, C. M. and LUTHERER, L. O. (1976). Effect of altitude exposure on thermoregulatory response of man to cold. *J. appl. Physiol.* **41,** 848–858.

BLATTEIS, C. M. and SMITH, K. A. (1979). Hypothalamic sensitivity to leucocytic pyrogen of adult and new-born guinea-pigs. *J. Physiol. (Lond.)* **296,** 177–192.

BLATTEIS, C. M., BILLMEIER, G. and GILBERT, T. (1973). Thermoregulation of phenylketonuric children. *Fed. Proc.* **32,** 408.

BLEICHERT, A., BEHLING, K., SCARPERI, M. and SCARPERI, S. (1973). Thermoregulatory behavior of man during rest and exercise. *Pflügers Arch.* **338,** 303–312.

BLIGH, J. (1961). Possible temperature-sensitive elements in or near the vena cava of sheep. *J. Physiol. (Lond.)* **159,** 85–86.

BLIGH, J. (1972). Neuronal models of mammalian temperature regulation. *In* "Essays on Temperature Regulation" (J. Bligh and R. E. Moore, Eds), pp. 105–120. North-Holland Publ. Comp., Amsterdam and London.

BLIGH, J. (1973). "Temperature Regulation in Mammals and other Vertebrates". North-Holland Publ. Comp., Amsterdam, London and New York.

BLIGH, J. (1974). Neuronal models of hypothalamic temperature regulation. *In* "Recent Studies of Hypothalamic Function" (K. Lederis and K. E. Cooper, Eds), pp. 315–327. Karger, Basel.

BLIGH, J. (1975). Neurotransmitters in temperature regulation. *In* "Selected Topics in Environmental Biology" (B. Bhatia, G. S. Chhina and B. Singh, Eds), pp. 3–10. Pergamon Press, Oxford.

BLIGH, J. (1976). Temperature regulation. *In* "Environmental Physiology of Animals" (J. Bligh, J. L. Cloudsley-Thompson and A. G. MacDonald, Eds), pp. 415–430. Blackwell, Oxford.

BLIGH, J. (1978). Thermoregulation: what is regulated and how? *In* "New Trends in Thermal Physiology" (Y. Houdas and J. D. Guieu, Eds), pp. 1–10. Masson, Paris.

BLIGH, J. (1979). The central neurology of mammalian thermoregulation. *Neuroscience* **4,** 1213–1236.

BLIGH, J. and HENSEL, H. (1974). Modern theories on the location and function of the thermoregulatory centres in mammals including man. *In* "Progress in Biometeorology" (S. W. Tromp, Ed.), Div. A, Vol. 1, Part 1B, pp. 413–433. Swets and Zeitlinger, Amsterdam.

BLIX, A. S. and STEEN, J. B. (1979). Temperature regulation in newborn polar homeotherms. *Physiol. Rev.* **59,** 285–304.

BLIX, M. (1882–1883). Experimentela bidrag till lösning af frågan om hudnervernas specifika energi. I. *Upsala Läk.-Fören. Förh.* **18,** 87–102.

BODEY, A. S. (1978). Changing cold acclimatization patterns of men living in Antarctica. *Int. J. Biometeorol.* **22,** 163–176.

BOIVIE, J. J. G. and PERL, E. R. (1975). Neural substrates of somatic sensation. *In* "Neurophysiology (C. C. Hunt, Ed.), pp. 303–411. Butterworths, London, and University Park Press, Baltimore.

BOMAN, K. K. A. (1958). Elektrophysiologische Untersuchungen über die Thermoreceptoren der Gesichtshaut. *Acta physiol. scand.* **44,** Suppl. 149.

BOMAN, K., HENSEL, H. and WITT, I. (1957). Die Entladung der Kalt-receptoren bei äußerer Einwirkung von Kohlensäure. *Pflügers Arch. ges. Physiol.* **264,** 107–112.

BONNER, R. M., HARRISON, M. H., HALL, C. J. and EDWARDS, R. J. (1976). Effect of heat acclimatization on intravascular responses to acute heat stress in man. *J. appl. Physiol.* **41,** 708–713.

BORSOOK, D., LABURN, H. P., ROSENDORFF, C., WILLIES, G. H. and WOOLF, C. J. (1977). A dissociation between temperature regulation and fever in the rabbit. *J. Physiol. (Lond.)* **266,** 423–433.

BOULANT, J. A. (1974). The effect of firing rate on preoptic neuronal thermosensitivity. *J. Physiol. (Lond.)* **240,** 661–669.

BOULANT, J. A. and BIGNALL, K. E. (1973a). Determinants of hypothalamic neuronal thermosensitivity in ground squirrels and rats. *Amer. J. Physiol.* **225,** 306–310.

BOULANT, J. A. and BIGNALL, K. E. (1973b). Changes in thermosensitive characteristics of hypothalamic units over time. *Amer. J. Physiol.* **225,** 311–318.

BOULANT, J. A. and BIGNALL, K. E. (1973c). Hypothalamic neuronal responses to peripheral and deep-body temperatures. *Amer. J. Physiol.* **225,** 1371–1374.

BOULANT, J. A. and DEMIEVILLE, H. N. (1977). Responses of thermosensitive preoptic and septal neurons to hippocampal and brain stem stimulation. *J. Neurophysiol.* **40,** 1356–1368.

BOULANT, J. A. and GONZALEZ, R. R. (1977). The effect of skin temperature on the hypothalamic control of heat loss and heat production. *Brain Res.* **120,** 367–372.

BOULANT, J. A. and HARDY, J. D. (1974). The effect of spinal and skin temperatures on the firing rate and thermosensitivity of preoptic neurones. *J. Physiol. (Lond.)* **240,** 639–660.

BOWSHER, D. (1975). Diencephalic projections from the midbrain reticular formation. *Brain Res.* **95,** 211–220.

BRAUN, H. A., BADE, H., SCHÄFER, K. and HENSEL, H. (1978). Dynamic characteristics of cold fibre discharge in the cat. *Pflügers Arch.* **373,** R67.

BRAUN, H. A., BADE, H. and HENSEL, H. (1980). Static and dynamic discharge patterns of bursting cold fibers related to hypothetical receptor mechanisms. *Pflügers Arch.* **386,** 1–9.

BREARLEY, E. A. and KENSHALO, D. R. (1970). Behavioral measurements of the sensitivity of cat's upper lip to warm and cool stimuli. *J. comp. physiol. Psychol.* **70,** 1–4.

BREBNER, D. F. and KERSLAKE, D. MCK. (1961). The effect of cyclical heating on the front of the trunk on the forearm sweat rate. *J. Physiol. (Lond.)* **158,** 144–153.

BRODAL, A., WALBERG, F. and TABER, E. (1960). Raphe nuclei of the brain stem of the cat. III. Afferent connections. *J. comp. Neurol.* **114,** 261–282.

BROWN, A. C. (1965). Equations of heat distribution within the body. *Bull. math. Biophys.* **27,** 67–78.

Brown, A. C. and Brengelmann, G. L. (1970). The interaction of peripheral and central inputs in the temperature regulation system. In "Physiological and Behavioral Temperature Regulation" (J. D. Hardy, A. P. Gagge and J. A. J. Stolwijk, Eds), pp. 684–702. Thomas, Springfield, Ill.

Brown, A. G. (1973). Ascending and long spinal pathways: Dorsal columns, spinocervical tract and spinothalamic tract. In "Handbook of Sensory Physiology", Vol. 2 (A. Iggo, Ed.), pp. 315–338. Springer, Berlin, Heidelberg and New York.

Brown, P. B., Moraff, H. and Tapper, D. N. (1973). Functional organization of the cat's dorsal horn: spontaneous activity and central cell response to single impulses in single type I fibers. J. Neurophysiol. **36**, 827–839.

Brück, K. (1961). Temperature regulation in the newborn infant. Biol. Neonat. (Basel) **3**, 65–119.

Brück, K. (1970a). Non-shivering thermogenesis and brown adipose tissue in relation to age, and their integration in the thermoregulatory system. In "Brown Adipose Tissue" (O. Lindberg, Ed.), pp. 117–154. Amer. Elsevier Publ. Comp., New York.

Brück, K. (1970b). Heat production and temperature regulation. In "Physiology of the Perinatal Period" (U. Stave, Ed.), Vol. 1, pp. 493–557. Appleton-Century-Crofts Meredith Corp., New York.

Brück, K. (1971). Die Temperaturregelung des Neugeborenen. In "Handbuch für Kinderheilkunde" (H. Opitz and F. Schmid, Eds), Part 1, pp. 23–30. Springer, Berlin, Heidelberg and New York.

Brück, K. (1976). Cold adaptation in man. In "Regulation of Depressed Metabolism and Thermogenesis" (L. Janský and X. J. Musacchia, Eds), pp. 42–63. Thomas, Springfield, Ill.

Brück, K. (1978a). Thermoregulation: control mechanisms and neural processes. In "Temperature Regulation and Energy Metabolism in the Newborn" (J. C. Sinclair, Ed.), pp. 157–185. Grune and Stratton, New York.

Brück, K. (1978b). Heat production and temperature regulation. In "Perinatal Physiology" (U. Stave, Ed.), pp. 455–498. Plenum Publ. Corp., New York.

Brück, K. (1978c). Ontogenetic and adaptive adjustments in the thermoregulatory system. In "New Trends in Thermal Physiology" (Y. Houdas and J. D. Guieu, Eds), pp. 65–77. Masson, Paris, New York, Barcelona and Milan.

Brück, K. and Schwennicke, H. P. (1971). Interaction of superficial and hypothalamic thermosensitive structures in the control of non-shivering thermogenesis. Int. J. Biometeorol. **15**, 156–161.

Brück, K. and Wünnenberg, W. (1967). Die Steuerung des Kältezitterns beim Meerschweinchen. Pflügers Arch. ges. Physiol. **293**, 215–225.

Brück, K. and Wünnenberg, W. (1970). "Meshed" control of two effector systems: nonshivering and shivering thermogenesis. In "Physiological and Behavioral Temperature Regulation." (J. D. Hardy, A. P. Gagge and J. A. J. Stolwijk, Eds), pp. 562–580. Thomas, Springfield, Ill.

BRÜCK, K., WÜNNENBERG, W., GALLMEIER, H. and ZIEHM, B. (1971). A mode of thermal adaptation: shift of threshold temperatures for shivering and heat polypnea. *J. Physiol. (Paris)* **63**, 213–221.

BRÜCK, K., BAUM, E. and SCHWENNICKE, H. P. (1976). Cold-adaptive modifications in man induced by repeated short-term cold-exposures and during a 10-day and -night cold-exposure. *Pflügers Arch.* **363**, 125–133.

BRYCE-SMITH, R., COLES, D. R., COOPER, K. E., CRANSTON, W. I. and GOODALE, F. (1959). The effects of intravenous pyrogen upon the radiant heat induced vasodilatation in man. *J. Physiol. (Lond.)* **145**, 77–84.

BULLARD, R. W., BANERJEE, M. R., CHEN, F., ELIZONDO, R. and MACINTYRE, B. A. (1970). Skin temperature and thermoregulatory sweating: a control systems approach. *In* "Physiological and Behavioral Temperature Regulation" (J. D. Hardy, A. P. Gagge and J. A. J. Stolwijk, Eds), pp. 597–610. Thomas, Springfield, Ill.

BURGESS, P. R. and PERL, E. R. (1973). Cutaneous mechanoreceptors and nociceptors. *In* "Handbook of Sensory Physiology" (A. Iggo, Ed.), Vol. 2, pp. 29–78. Springer, Berlin, Heidelberg and New York.

BURTON, H. (1975). Responses of spinal cord neurons to systematic changes in hindlimb skin temperatures in cats and primates. *J. Neurophysiol.* **38**, 1060–1079.

BYKOV, K. M. (1959). "The Cerebral Cortex and the Internal Organs", pp. 248–270. Foreign Languages Publ. House, Moscow.

CABANAC, M. (1969). Plaisir ou déplaisir de la sensation thermique et homeothermie. *Physiol. Behav.* **4**, 359–364.

CABANAC, M. (1970). Interaction of cold and warm temperature signals in the brain stem. *In* "Physiological and Behavioral Temperature Regulation" (J. D. Hardy, A. P. Gagge and J. A. J. Stolwijk, Eds), pp. 549–561. Thomas, Springfield, Ill.

CABANAC, M. (1972). Thermoregulatory behavior. *In* "Essays on Temperature Regulation" (J. Bligh and R. E. Moore, Eds), pp. 19–32. North-Holland Publ. Comp., Amsterdam and London.

CABANAC, M. (1975). Temperature regulation. *Ann. Rev. Physiol.* **37**, 415–439.

CABANAC, M. (1979). Les signaux physiologiques du comfort thermique. *In* "Thermal Comfort" (J. Durand and J. Raynaud, Eds), Vol. 75, pp. 27–38. Editions INSERM, Paris.

CABANAC, M. and CAPUTA, M. (1979). Open loop increase in trunk temperatures produced by face cooling in working humans. *J. Physiol. (Lond.)* **289**, 163–174.

CABANAC, M. and HARDY, J. D. (1969). Réponses unitaires et thermorégulatrices lors de réchauffements et réfroidissements localisés de la région préoptique et du mésencéphale chez le lapin. *J. Physiol. (Paris)* **61**, 331–347.

CABANAC, M. and MASSONNET, B. (1974). Temperature regulation during fever: change of set point or change of gain? A tentative answer from a behavioural study in man. *J. Physiol. (Lond.)* **238**, 561–568.

CABANAC, M. and MASSONNET, B. (1977). Thermoregulatory responses as a function of core temperature in humans. *J. Physiol. (Lond.)* **265**, 587–596.

CABANAC, M., CHATONNET, J. and PHILIPOT, R. (1965). Les conditions de températures cérébrale et cutanée moyennes pour l'apparition du frisson thermique chez le chien. *C. r. Soc. Biol. (Paris)* **260**, 680–683.

CABANAC, M., DUCLAUX, R. and Chatonnet, J. (1966). Influence d'une élévation passive de la température interne sur le comportement thermo-régulateur du chien. *J. Physiol. (Paris)* **58**, 214.

CABANAC, M., STOLWIJK, J. A. J. and HARDY, J. D. (1968). Effect of temperature and pyrogens on single-unit activity in the rabbit's brain stem. *J. appl. Physiol.* **24**, 645–652.

CABANAC, M., DUCLAUX, R. and GILLET, A. (1970). Thermoregulation comportementale chez le chien: effets de la fièvre et de la thyroxine. *Physiol. Behav.* **5**, 697–704.

CABANAC, M., CUNNINGHAM, D. J. and STOLWIJK, J. A. J. (1971). Thermoregulatory set point during exercise: a behavioral approach. *J. comp. physiol. Psychol.* **76**, 94–102.

CABANAC, M., HILDEBRANDT, G., MASSONNET, B. and STREMPEL, H. (1976). A study of the nycthemeral cycle of behavioural temperature regulation in man *J. Physiol. (Lond.)* **257**, 275–291.

CALVERT, D. T. and FINDLAY, J. D. (1975). Localization of the effective thermosensitive site in the preoptic region of the ox. *J. appl. Physiol.* **39**, 702–706.

CARLISLE, H. J. (1966). Behavioral significance of hypothalamic tempera-ture-sensitive cells. *Nature (Lond.)* **209**, 1324–1325.

CARLISLE, H. J. (1968a). Peripheral thermal stimulation and thermoregula-tory behavior. *J. comp. physiol. Psychol.* **66**, 507–510.

CARLISLE, H. J. (1968b). Initiation of behavioral responding for heat in a cold environment. *Physiol. Behav.* **3**, 827–830.

CARLISLE, H. J. (1969). Effect of preoptic and anterior hypothalamic lesions on behavioral thermoregulation in the cold. *J. comp. physiol. Psychol.* **69**, 391–402.

CARLISLE, H. J. (1970). Thermal reinforcement and temperature regulation. *In* "Animal Psychophysics: The Design and Conduct of Sensory Experi-ments" (W. C. Stebbins, Ed.), pp. 211–229. Appleton-Century-Crofts Meredith Corp., New York.

CARLISLE, H. J. and INGRAM, D. L. (1973a). The influence of body core temperature and peripheral temperatures on oxygen consumption in the pig. *J. Physiol. (Lond.)* **231**, 341–352.

CARLISLE, H. J. and INGRAM, D. L. (1973b). The effects of heating and cooling the spinal cord and hypothalamus on thermoregulatory behaviour in the pig. *J. Physiol. (Lond.)* **231**, 353–364.

CARLSSON, A., FALCK, B. and HILLARP, N.-A. (1962). Cellular localization of brain monoamines. *Acta physiol. scand.* **56**, Suppl. 196.

CARLTON, P. L. and MARKS, R. A. (1958). Cold exposure and heat reinforced operant behaviour. *Science* **128**, 1344.

CARSTENS, E., YOKOTA, T. and ZIMMERMANN, M. (1979). Inhibition of spinal neuronal responses to noxious skin heating by stimulation of mesencephalic periaqueductal gray in the cat. *J. Neurophysiol.* **42**, 558–568.

CERVERO, F., IGGO, A. and MOLONY, V. (1977). Responses of spinocervical tract neurones to noxious stimulation of the skin. *J. Physiol. (Lond.)* **267**, 537–558.

CHAFFEE, R. R. J. and ROBERTS, J. C. (1971). Temperature acclimation in birds and mammals. *Ann. Rev. Physiol.* **33**, 155–202.

CHAI, C. Y. and LIN, M. T. (1972). Effects of heating and cooling the spinal cord and medulla oblongata on thermoregulation in monkeys. *J. Physiol. (Lond.)* **225**, 297–308.

CHAI, C. Y. and LIN, M. T. (1973). Effects of thermal stimulation of medulla oblongata and spinal cord on decerebrate rabbits. *J. Physiol. (Lond.)* **234**, 409–419.

CHAMBERS, M. R., ANDRES, K. H., VON DÜRING, M. and IGGO, A. (1972). The structure and function of the slowly adapting type II mechanoreceptor in hairy skin. *Quart. J. exp. Physiol.* **57**. 417–445.

CHATONNET, J. and CABANAC, M. (1965). The perception of thermal comfort. *Int. J. Biometeorol.* **9**, 183–193.

CHATONNET, J., CABANAC, M. and MOTTAZ, M. (1964). Les conditions de températures cérébrale et cutanée moyenne pour l'apparition de la polypnée thermique chez le chien. *C. r. Soc. Biol. (Paris)* **158**, 1354–1356.

CHATONNET, J., CABANAC, M. and SATTA, S. (1964). Evolution de la sensation thermique au cours de l'exposition au froid chez l'homme. *C. r. Soc. Biol. (Paris)* **158**, 305–307.

CHATONNET, J., THIERS, H., CABANAC, M. and PASQUIER, J. (1966). Sur l'origine de l'impression consciente de comfort thermique. *Lyon méd.* **50**, 1387–1392.

CHATT, A. B. and KENSHALO, D. R. (1977). Cerebral evoked responses to skin warming recorded from human scalp. *Exp. Brain Res.* **28**, 449–455.

CHATT, A. B. and KENSHALO, D. R. (1979). The afferent fiber population mediating the thermal evoked response to skin cooling in man. *Exp. Neurol.* **64**, 146–154.

CHRISTENSEN, B. N. and PERL, E. R. (1970). Spinal neurons specifically excited by noxious or thermal stimuli: marginal zone of the dorsal horn. *J. Neurophysiol.* **33**, 293–307.

CLARK, W. G. and LIPTON, J. M. (1974). Complementary lowering of the behavioural and physiological thermoregulatory setpoint by tetrodotoxin and saxitoxin in the cat. *J. Physiol. (Lond.)* **238**, 181–191.

CLOUGH, D. P. and JESSEN, C. (1974). The role of spinal thermosensitive structures in the respiratory heat loss during exercise. *Pflügers Arch.* **347**, 235–248.

CLOUGH, D. P., DARLING, K. F., FINDLAY, J. D. and THOMPSON, G. E. (1973). Cold sensitivity in the spinal cord of sheep. *Pflügers Arch.* **342**, 137–144.

COLIN, J. and HOUDAS, Y. (1968). Déterminisme du déclenchement de la sudation thermique chez l'homme. *J. Physiol. (Paris)* **60**, 5–31.

COLIN, J., TIMBAL, J., BOUTELIER, CH., HOUDAS, Y. and SIFFRE, M. (1968). Rhythm of the rectal temperature during a 6-month free-running experiment. *J. appl. Physiol.* **25,** 170–176.

COLLINS, K. J. (1979). Hyperthermia and thermal responsiveness in the elderly. In "Indoor Climate" (P. O. Fanger and O. Valbjørn, Eds), pp. 819–833. Danish Building Res. Inst., Copenhagen.

COLLINS, K. J. and WEINER, J. S. (1968). Endocrinological aspects of exposure to high environmental temperatures. *Physiol. Rev.* **48,** 785–839.

COLLINS, K. J., DORÉ, C., EXTON-SMITH, A. N., FOX, R. H., MACDONALD, I. C. and WOODWARD, P. M. (1977). Accidental hypothermia and impaired temperature homeostasis in the elderly. *Brit. med. J.* **1,** 353–356.

CONROY, R. T. W. L. and MILLS, J. N. (1970). "Human Circadian Rhythms". Churchill, London.

COOPER, K. E. (1972). The body temperature "set-point" in fever. In "Essays on Temperature Regulation" (J. Bligh and R. E. Moore, Eds), pp. 149–162. North-Holland Publ. Comp., Amsterdam and London.

COOPER, K. E., CRANSTON, W. I. and SNELL, E. S. (1964a). Temperature regulation during fever in man. *Clin. Sci.* **27,** 345–356.

COOPER, K. E., JOHNSON, R. H. and SPALDING, J. M. K. (1964b). Thermoregulatory reactions following intravenous pyrogens in a subject with complete transection of the cervical cord. *J. Physiol. (Lond.)* **171,** 55P–56P.

COOPER, K. E., CRANSTON, W. I. and HONOUR, A. J. (1965). Effects of intraventricular and intrahypothalamic injection of noradrenaline and 5-HT on body temperature in conscious rabbits. *J. Physiol. (Lond.)* **181,** 852–864.

COOPER, K. E., CRANSTON, W. I. and HONOUR, A. J. (1967). Observations on the site and mode of action of pyrogens in the rabbit brain. *J. Physiol. (Lond.)* **191,** 325–337.

COOPER, K. E., PITTMAN, Q. J. and VEALE, W. L. (1975). Observations on the development of the 'fever' mechanism in the fetus and newborn. In "Temperature Regulation and Drug Action" (J. Lomax, E. Schönbaum and J. Jacob, Eds), pp. 43–50. Karger, Basel.

COOPER, K. E., JONES, D. L., PITTMAN, Q. J. and VEALE, W. L. (1976a). The effect of noradrenaline, injected into the hypothalamus, on thermoregulation in the cat. *J. Physiol. (Lond.)* **261,** 211–222.

COOPER, K. E., JONES, D. L., PITTMAN, Q. J. and VEALE, W. L. (1976a). The effect of noradrenaline, injected into the hypothalamus, on thermoregulation in the cat. *J. Physiol. (Lond.)* **261,** 211–222.

COOPER, K. E., PRESTON, E. and VEALE, W. L. (1976b). Effects of atropine, injected into a lateral cerebral ventricle of the rabbit, on fevers due to intravenous leucocyte pyrogen and hypothalamic and intraventricular injections of prostaglandin E_1. *J. Physiol. (Lond.)* **254,** 729–741.

CORBIT, J. D. (1969). Behavioral regulation of hypothalamic temperature. *Science* **166,** 256–258.

CORBIT, J. D. (1970). Behavioral regulation of body temperature. In "Physiological and Behavioral Temperature Regulation" (J. D. Hardy, A. P. Gagge and J. A. J. Stolwijk, Eds), pp. 777–801. Thomas, Springfield, Ill.

CORBIT, J. D. (1973). Voluntary control of hypothalamic temperature. J. comp. physiol. Psychol. **83**, 394–411.

CORBIT, J. D. and ERNITS, T. (1974). Specific preference for hypothalamic cooling. J. comp. physiol. Psychol. **86**, 24–27.

CORMARÈCHE-LEYDIER, M. and CABANAC, M. (1973). Influence de stimulations thermiques de la moëlle épinière sur le comportement thermorégulateur du chien. Pflügers Arch. **341**, 313–324.

CORMARÈCHE-LEYDIER, M., BANET, M., HENSEL, H. and CABANAC, M. (1977). Thermoregulatory behavior after repetitive cooling of the preoptic area and of the spinal cord in the rat. Pflügers Arch. **369**, 17–20.

CRAIG, W. S. (1963). The early detection of pyrexia in the newborn. Arch. Dis. Childh. **38**, 29–39.

CRANSTON, W. I. and RAWLINS, M. D. (1972). Effects of intracerebral micro-injection of sodium salicylate on temperature regulation in the rabbit. J. Physiol. (Lond.) **222**, 257–266.

CRANSTON, W. I., HELLON, R. F., LUFF, R. H. and RAWLINS, M. D. (1972). Hypothalamic endogenous noradrenaline and thermoregulation in the cat and rabbit. J. Physiol. (Lond.) **223**, 59–67.

CRANSTON, W. I., HELLON, R. F. and MITCHELL, D. (1975). A dissociation between fever and prostaglandin concentration in cerebrospinal fluid. J. Physiol. (Lond.) **253**, 583–592.

CRANSTON, W. I., DUFF, G. W., HELLON, R. F. and MITCHELL, D. (1976a). Thermoregulation in rabbits during fever. J. Physiol. (Lond.) **257**, 767–777.

CRANSTON, W. I., DUFF, G. W., HELLON, R. F., MITCHELL, D. and TOWNSEND, Y. (1976b). Evidence that brain prostaglandin synthesis is not essential in fever. J. Physiol. (Lond.) **259**, 239–249.

CRANSTON, W. I., HELLON, R. F. and TOWNSEND, Y. (1978). Thermal stimulation of intra-abdominal veins in conscious rabbits. J. Physiol. (Lond.) **277**, 49–52.

CRAWSHAW, L. I. and STITT, J. T. (1975). Behavioural and autonomic induction of prostaglandin E₁ fever in squirrel monkeys. J. Physiol. (Lond.) **244**, 197–206.

CRAWSHAW, L. I., NADEL, E. R., STOLWIJK, J. A. J. and STAMFORD, B. A. (1975). Effect of local cooling on sweating rate and cold sensation. Pflügers Arch. **354**, 19–27.

CROCKFORD, G. W., FOSTER, K. P. and HASPINEALL, J. (1971). Factors determining the sweating threshold. Int. J. Biometeorol. **15**, 286–291.

CROZE, S., DUCLAUX, R. and KENSHALO, D. R. (1976). The thermal sensitivity of the polymodal nociceptors in the monkey. J. Physiol. (Lond.) **263**, 539–562.

CUNNINGHAM, D. J. and CABANAC, M. (1971). Evidence from behavioral thermoregulatory responses of a shift in setpoint temperature related to the menstrual cycle. J. Physiol. (Paris) **63**, 236–238.

CUNNINGHAM, D. J., STOLWIJK, J. A. J., MURAKAMI, N. and HARDY, J. D. (1967). Responses of neurons in the preoptic area to temperature, serotonin and epinephrine. *Amer. J. Physiol.* **213,** 1570–1581.

CUNNINGHAM, D. J., STOLWIJK, J. A. J. and WENGER, C. B. (1978). Comparative thermoregulatory responses of resting men and women. *J. appl. Physiol.* **45,** 908–915.

DAHLSTRÖM, A. and FUXE, K. (1964). Evidence for the existence of monoamine-containing neurons in the central nervous system. I. Demonstration of monoamines in the cell bodies of brain stem neurons. *Acta physiol. scand.* **62,** Suppl. 232.

DALI, J., FUXE, K. and JONSSON, G. (1974). 5,7-Dihydroxytryptamine as a tool for morphological and functional analysis of central 5-hydroxytryptamine neurons. *Res. Comn. chem. Pathol. Pharmac.* **7,** 175–187.

DARIAN-SMITH, I. (1973). The trigeminal system. *In* "Handbook of Sensory Physiology", Vol. 2 (A. Iggo, Ed.), pp. 274–314. Springer, Berlin, Heidelberg and New York.

DARIAN-SMITH, I. and DYKES, R. W. (1971). Peripheral neural mechanisms of thermal sensations. *In* "Oral-Facial Sensory and Motor Mechanisms" (R. Dubner and Y. Kawamura, Eds), pp. 7–22. Appleton-Century Crofts Meredith Corp., New York.

DARIAN-SMITH, I., JOHNSON, K. O. and DYKES, R. (1973). "Cold" fiber population innervating palmar and digital skin of the monkey: responses to cooling pulses. *J. Neurophysiol.* **36,** 325–346.

DARIAN-SMITH, I., JOHNSON, K. O. and LA MOTTE, C. (1975). Peripheral neural determinants in the sensing of changes in skin temperature. *In* "The Somatosensory System" (H. H. Kornhuber, Ed.), pp. 23–37. Thieme, Stuttgart.

DARIAN-SMITH, I., JOHNSON, K. O., LaMOTTE, C., KENINS, P., SHIGENAGA, Y. and MING, V. C. (1979a). Coding of incremental changes in skin temperature by single warm fibers in the monkey. *J. Neurophysiol.* **42,** 1316–1331.

DARIAN-SMITH, I., JOHNSON, K. O., LaMOTTE, C., SHIGENAGA, Y., KENINS, P. and CHAMPNESS, P. (1979b). Warm fibers innervating palmar and digital skin of the monkey: responses to thermal stimuli. *J. Neurophysiol.* **42,** 1297–1315.

DAVIS, T. R. A. (1961). Chamber cold acclimatization in man. *J. appl. Physiol.* **16,** 1011–1015.

DHUMAL, V. R. and GULATI, O. D. (1973). Effect on body temperature in dogs of perfusion of cerebral ventricles with artificial CSF deficient in calcium or containing excess of sodium or calcium. *Brit. J. Pharmacol.* **49,** 699–701.

DINARELLO, C. A. (1980) Endogenous pyrogens. *In* "Fever" (J. M. Lipton, Ed.), pp. 1–9. Raven Press, New York.

DINARELLO, C. A. and WOLFF, S. M. (1978). Pathogenesis of fever in man. *New Engl. J. Med.* **298,** 607–612.

DINARELLO, C. A., RENFER, L. and WOLFF, S. M. (1977). Human leukocytic pyrogen: purification and development of a radioimmunoassay. *Proc. nat. Acad. Sci. (Wash.)* **74**, 4624.

DODT, E. (1953). The behaviour of thermoreceptors at low and high temperatures with special reference to Ebbecke's temperature phenomena. *Acta physiol. scand.* **27**, 295–314.

DODT, E. (1956). Die Aktivität der Thermoreceptoren bei nicht-thermischen Reizen bekannter thermoregulatorischer Wirkung. *Pflügers Arch. ges. Physiol.* **263**, 188–200.

DODT, E. and ZOTTERMAN, Y. (1952). The discharge of specific cold fibres at high temperatures. (The paradoxical cold.) *Acta physiol. scand.* **26**, 358–365.

DODT, E., SKOUBY, A. P. and ZOTTERMAN, Y. (1953). The effect of cholinergic substances on the discharges from thermal receptors. *Acta physiol. scand.* **28**, 101–114.

DOI, K. and KUROSHIMA, A. (1979). Lasting effect of infantile cold experience on cold tolerance in adult rats. *Jap. J. Physiol.* **29**, 139–150.

DOI, K., OHNO, T., KURAHASHI, M. and KUROSHIMA, A. (1979). Thermoregulatory nonshivering thermogenesis in men, with special reference to lipid metabolism. *Jap. J. Physiol.* **29**, 359–372.

DOSTROVSKY, J. O. and HELLON, R. F. (1978). The representation of facial temperature in the caudal trigeminal nucleus of the cat. *J. Physiol. (Lond.)* **277**, 29–47.

DOWNEY, J. A., MILLER, J. M. and DARLING, R. C. (1969). Thermoregulatory responses to deep and superficial cooling in spinal man. *J. appl. Physiol.* **27**, 209–212.

DOWNEY, J. A., HUCKABA, C. E. and DARLING, R. C. (1971). The effect of skin and central cooling on human thermoregulation. *Int. J. Biometeorol.* **15**, 171–175.

DOWNEY, J. A., HUCKABA, C. E., KELLEY, P. S., TAM, H. S., DARLING, R. C. and CHEH, H. Y. (1976). Sweating responses to central and peripheral heating in spinal man. *J. appl. Physiol.* **40**, 701–706.

DRESSENDORFER, R. H., SMITH, R. M., BAKER, D. G. and HONG, S. K. (1977). Cold tolerance of long-distance runners and swimmers in Hawaii. *Int. J. Biometeorol.* **21**, 51–63.

DUBNER, R., SUMINO, R. and WOOD, W. I. (1975). A peripheral "cold" fiber population responsive to innocuous and noxious thermal stimuli applied to monkey's face. *J. Neurophysiol.* **38**, 1373–1389.

DUCLAUX, R. and KENSHALO, D. R. (1980). Response characteristics of cutaneous warm receptors in the monkey. *J. Neurophysiol.* **43**, 1–15.

DUCLAUX, R., FRANZEN, O., CHATT, A. B., KENSHALO, D. R. and STOWELL, H. (1974). Responses recorded from human scalp evoked by cutaneous thermal stimulation. *Brain Res.* **78**, 279–290.

DUCLAUX, R., SCHÄFER, K. and HENSEL, H. (1980). Response of cold receptors to low skin temperatures in the nose of the cat. *J. Neurophysiol.* **43**, 1571–1577.

DURAND, J. and RAYNAUD, J. (Eds) (1979). "Thermal Comfort". Editions INSERM, Paris.

DYKES, R. W. (1975). Coding of steady and transient temperatures by cutaneous 'cold' fibers serving the hand of monkeys. *Brain Res.* **98,** 485–500.

DYMNIKOVA, L. P. (1973). Activity in the posterior hypothalamus neurons during brain and skin temperature changes in unanesthetized rabbits. *Neurophysiologia (Kiew)* **5,** 490–496.

DYMNIKOVA, L. P., ZAKHARZHEVSKAYA, N. P. and IVANOV, K. P. (1973). On afferent projections of the thermoregulation center. *Sechenov physiol. J. USSR* **59,** 156–163.

EAGAN, C. J. (1963). Local vascular adaptations to cold in man. *Fed. Proc.* **22,** 947–951.

EASTMAN, C. J., EKINS, R. P., LEITH, I. M. and WILLIAMS, E. S. (1974). Thyroid hormone response to prolonged cold exposure in man. *J. Physiol. (Lond.)* **241,** 175–181.

EBBECKE, U. (1917). Über die Temperaturempfindungen in ihrer Abhängigkeit von der Hautdurchblutung und von den Reflexzentren. *Pflügers Arch. ges. Physiol.* **169,** 395–462.

EBBECKE, U. (1948). Schüttelfrost in Kälte, Fieber und Affekt. *Klin. Wschr.* **26,** 609–613.

EGGERS, H. J. (1971). Bedeutung von Fieber und Hyperthermie für den Verlauf von Virusinfektionen. *Z. phys. Med.* **2,** 70–77.

EIDE, R. (1971). Spatial specificity of local habituation to cold. *Acta physiol. scand.* **82,** 433–438.

EIDE, R. (1976). Physiological and behavioral reactions to repeated tail cooling in the white rat. *J. appl. Physiol.* **41,** 292–294.

EISENMAN, J. S. (1969). Pyrogen-induced changes in the thermosensitivity of septal and preoptic neurons. *Amer. J. Physiol.* **216,** 330–334.

EISENMAN, J. S. (1970). The action of bacterial pyrogen on thermoresponsive neurons. *In* "Physiological and Behavioral Temperature Regulation" (J. D. Hardy, A. P. Gagge and J. A. J. Stolwijk, Eds), pp. 507–518. Thomas, Springfield, Ill.

EISENMAN, J. S. (1972). Unit activity studies of thermoresponsive neurons. *In* "Essays on Temperature Regulation" (J. Bligh and R. E. Moore, Eds), pp. 55–69. North-Holland Publ. Comp., Amsterdam and London.

EISENMAN, J. S. (1974a). Unit studies of brainstem projections to the preoptic area and hypothalamus. *In* "Recent Studies in Hypothalamic Function" (K. Lederis and K. E. Cooper, Eds), pp. 328–340. Karger, Basel.

EISENMAN, J. S. (1974b). Depression of preoptic thermosensitivity by bacterial pyrogen in rabbits. *Amer. J. Physiol.* **227,** 1067–1073.

ELIZONDO, R. S. and BULLARD, R. W. (1971). Local determinants of sweating and the assessment of the "set point". *Int. J. Biometeorol.* **15,** 273–280.

ELIZONDO, R. S., BANERJEE, M. and BULLARD, R. W. (1972). Effect of local heating and arterial occlusion on sweat electrolyte content. *J. appl. Physiol.* **32,** 1–6.

EPSTEIN, H. C., HOCHWALD, A. and ASHE, R. (1951). Salmonella infections of the newborn infant. *J. Pediat.* **38,** 723–731.

ERICKSON, R. P. and POULOS, D. A. (1973). On the qualitative aspect of the temperature sense. *Brain Res.* **61**, 107–112.

ERLANGER, J. and GASSER, H. S. (1937). "Electrical Signs of Nervous Activity". Univ. of Philadelphia Press, Philadelphia, Pa.

EULER, U. S. VON, LINDER, E. and MYRIN, S. O. (1943). Über die fiebererregende Wirkung des Adrenalins. *Acta physiol. scand.* **5**, 85–96.

FANGER, P. O. (1970). "Thermal Comfort". Danish Technical Press, Copenhagen.

FANGER, P. O. (1973a). "Thermal Comfort". McGraw-Hill Book Comp., New York.

FANGER, P. O. (1973b). Thermal environments preferred by man. *Build int.* **6**, 127–141.

FANGER, P. O. (1973c). Assessment of man's thermal comfort in practice. *Brit. J. industr. Med.* **30**, 313–324.

FANGER, P. O. (1973d). The influence of age, sex, adaptation, season and circadian rhythm on thermal comfort criteria for man. *Bull. Inst. int. Froid* **2**, 91–97.

FANGER, P. O. (1976). Energy conservation and human comfort. *In* "Ventilation and Air Conditioning". 8th Conference on Heating and Ventilating Budapest, Hungary, 1976. Conference Proceedings, pp. 1–30.

FANGER, P. O. (1977). Thermal comfort in indoor environments. *In* "Thermal Analysis—Human Comfort—Indoor Environments" (B. W. Mangum and J. E. Hill, Eds), p. 7. National Bureau Stand. US Publ. 491, Washington, D.C.

FANGER, P. O. (1979). Thermal discomfort caused by radiant asymmetry, local air velocities, warm or cold floors, and vertical air temperature gradients. *In* "Thermal Comfort" (J. Durand and J. Raynaud, Eds), Vol. 75, pp. 145–152. Editions INSERM, Paris.

FANGER, P. O. and VALBJØRN, O. (Eds) (1979). "Indoor Climate". Danish Building Res. Inst., Copenhagen.

FANGER, P. O., HØJBJERRE, J. and THOMSEN, J. O. B. (1974a). Thermal comfort conditions in the morning and in the evening. *Int. J. Biometeorol.* **18**, 16–22.

FANGER, P. O., ÖSTBERG, O., MCK. NICHOLL, A. G., BREUM, N. O. and JERKING, E. (1974b). Thermal comfort conditions during day and night. *Europ. J. appl. Physiol.* **33**, 255–263.

FANGER, P. O., HØJBJERRE, J. and THOMSEN, J. O. B. (1977). Can winter swimming cause people to prefer lower room temperatures? *Int. J. Biometeorol.* **21**, 44–50.

FELDBERG, W. and GUPTA, K. P. (1973). Pyrogen fever and prostaglandin-like activity in cerebrospinal fluid. *J. Physiol. (Lond.)* **228**, 41–53.

FELDBERG, W. and MYERS, R. D. (1963). A new concept of temperature regulation by amines in the hypothalamus. *Nature (Lond.)* **200**, 1325.

FELDBERG, W. and MYERS, R. D. (1964). Effects on temperature of amines injected into the cerebral ventricles. A new concept of temperature regulation. *J. Physiol. (Lond.)* **173**, 226–236.

FELDBERG, W. and MYERS, R. D. (1965). Changes in temperature produced by microinjections of amines in the anterior hypothalamus of cats. *J. Physiol. (Lond.)* **177**, 239–245.

FELDBERG, W. and MYERS, R. D. (1966). Appearance of 5-hydroxytryptamine and an unidentified pharmacologically active lipid acid in effluent from perfused cerebral ventricles. *J. Physiol. (Lond.)* **184**, 837–855.

FELDBERG, W. and SAXENA, P. N. (1971a). Effects of adrenoceptor blocking agents on body temperature. *Brit. J. Pharmacol.* **43**, 543–554.

FELDBERG, W. and SAXENA, P. N. (1971b). Fever produced by prostaglandin E_1. *J. Physiol. (Lond.)* **217**, 547–556.

FELDBERG, W. and SAXENA, P. N. (1971c). Further studies on prostaglandin E_1 fever in cats. *J. Physiol. (Lond.)* **219**, 739–745.

FELDBERG, W., HELLON, R. F. and LOTTI, V. J. (1967). Temperature effects produced in dogs and monkeys by injections of monoamines and related substances into the third ventricle. *J. Physiol. (Lond.)* **191**, 501–515.

FELDBERG, W., GUPTA, K. P., MILTON, A. S. and WENDLANDT, S. (1973). Effect of pyrogen and antipyretics on prostaglandin activity in cisternal c.s.f. of unanaesthetized cats. *J. Physiol. (Lond.)* **234**, 279–303.

FERREIRA, S. H., MONCADA, S. and VANE, J. R. (1971). Indomethacin and aspirin abolish prostaglandin release from the spleen. *Nature (New Biol.)* **231**, 237.

FINDLAY, A. L. R. and HAYWARD, J. N. (1969). Spontaneous activity of single neurones in hypothalamus of rabbits during sleep and waking. *J. Physiol. (Lond.)* **201**, 237–258.

FLOWER, R. J. and VANE, J. R. (1972). Inhibition of prostaglandin synthesis in brain explains the antipyretic activity of paracetamol (4-acetamidophenol). *Nature (Lond.)* **240**, 410–411.

FORD, D. M. (1974). A diencephalic island for the study of thermally-responsive neurones in the cat's hypothalamus. *J. Physiol. (Lond.)* **239**, 67P–68P.

FORSLING, M. L., INGRAM, D. L. and STANIER, M. W. (1976). Effects of various ambient temperatures and of heating and cooling the hypothalamus and spinal cord on antidiuretic hormone secretion and urinary osmolality in pigs. *J. Physiol. (Lond.)* **257**, 673–686.

FOSTER, K. G. (1971). Factors affecting the quantitative response of human eccrine sweat glands to intradermal injections of acetylcholine and methacholine. *J. Physiol. (Lond.)* **213**, 277–290.

FOSTER, K. G. and WEINER, J. S. (1970). Effects of cholinergic and adrenergic blocking agents on the activity of the eccrine sweat glands. *J. Physiol. (Lond.)* **210**, 883–895.

FOSTER, K. G., HEY, E. N. and KATZ, G. (1969). The response of the sweat glands of the new-born baby to thermal stimuli and to intradermal acetylcholine. *J. Physiol. (Lond.)* **203**, 13–29.

FOX, R. H., GOLDSMITH, R., KIDD, D. J. and LEWIS, H. E. (1963). Blood flow and other thermoregulatory changes with acclimatization to heat. *J. Physiol. (Lond.)* **166**, 548–562.

Fox, R. H., Löfstedt, B. E., Woodward, P. M., Ericksson, E. and Werkstrom, B. (1969). Comparison of thermoregulatory function in men and women. *J. appl. Physiol.* **26**, 444–453.

Fox, R. H., Woodward, P. M., Exton-Smith, A. N., Green, M. F. and Donnison, D. V. (1973). Body temperatures in the elderly: a national study of physiological, social and environmental conditions. *Brit. med. J.* **1**, 200–206.

Freeman, W. J. and Davis, D. D. (1959). Effect on cats of conductive hypothalamic cooling. *Amer. J. Physiol.* **197**, 145–148.

Freeman, M. E., Crissman, J. K. Jr, Louw, G. N., Butcker, R. L. and Inskeep, F. K. (1970). Thermogenic action of progesterone in the rat. *Endocrinology* **86**, 717–720.

Frens, J. (1971). Central synaptic interference and experimental fever. *Int. J. Biometeorol.* **15**, 313–315.

Frey, M. von (1895). Beiträge zur Sinnesphysiologie der Haut. III. *Ber. sächs. Ges. (Akad.) Wiss.* **47**, 166–184.

Frey, M. von (1910). Physiologie der Sinnesorgane der menschlichen Haut. *Ergebn. Physiol.* **9**, 351–368.

Fruhstorfer, H. and Hensel, H. (1973). Thermal cutaneous afferents in the trigeminal nucleus of the cat. *Naturwissenschaften* **60**, 209.

Fruhstorfer, H., Guth, H. and Pfaff, U. (1972). Thermal reaction time as a function of stimulation site. *Pflügers Arch.* **335**, R49.

Fruhstorfer, H., Guth, H. and Pfaff, U. (1973). Cortical responses evoked by thermal stimuli in man. 3rd Int. Congr. Event Related Slow Potentials of the Brain, 1973.

Fruhstorfer, H., Zenz, M., Nolte, H. and Hensel, H. (1974). Dissociated loss of cold and warm sensibility during regional anaesthesia. *Pflügers Arch.* **349**, 73–82.

Fuller, C. A., Sulzman, F. M. and Moore-Ede, M. C. (1977). The effect of suprachiasmatic nucleus lesions on circadian rhythms in the squirrel monkey (Saimiri sciureus). *Soc. Neuroscience* Abstr. III, 162.

Fuller, C. A., Sulzman, F. M. and Moore-Ede, M. C. (1978). Thermoregulation is impaired in an environment without circadian time cues. *Science* **199**, 794–796.

Fuller, C. A., Sulzman, F. M. and Moore-Ede, M. C. (1979). Circadian control of thermoregulation in the squirrel monkey, Saimiri sciureus. *Amer. J. Physiol.* **236**, R153–R161.

Fuller, J. H. (1975). Brain stem reticular units: some properties of the course and origin of the ascending trajectory. *Brain Res.* **83**, 349–367.

Fuxe, K. (1965). The distribution of monoamine terminals in the central nervous system. *Acta physiol. scand.* **64**, Suppl. 247, 37–85.

Gagge, A. P. (1979). Introduction to thermal comfort. *In* "Thermal Comfort" (J. Durand and J. Raynaud, Eds), pp. 11–24. Editions INSERM, Paris.

GAGGE, A. P. and NISHI, Y. (1977). Heat exchange between human skin surface and thermal environment. *In* "Handbook of Physiology" (D. H. K. Lee, Ed.), Sect. 9, pp. 69–92. Amer. Physiol. Soc., Bethesda, Md.

GAGGE, A. P., STOLWIJK, J. A. J. and HARDY, J. D. (1967). Comfort and thermal sensations and associated physiological responses at various ambient temperatures. *Environ. Res.* **1**, 1–20.

GAGGE, A. P., STOLWIJK, J. A. J. and SALTIN, B. (1969). Comfort and thermal sensations and associated physiological responses during exercise at various ambient temperatures. *Environ. Res.* **2**, 209–229.

GALE, C. C., MATTHEWS, M. and YOUNG, J. (1970). Behavioral thermoregulatory responses to hypothalamic cooling and warming in baboons. *Physiol. Behav.* **5**, 1–6.

GALLEGO, R., EYZAGUIRRE, C. and MONTI-BLOCH, L. (1979). Thermal and osmotic responses of arterial receptors. *J. Neurophysiol.* **42**, 665–680.

GAUTHERIE, M. (1971). Mechanism of local skin thermoregulation in man essentially controlled by a cooperative biosynthesis of bradykinin. *J. Physiol. (Paris)* **63**, 251–253.

GIBIŃSKI, K., GIEC, L., ZMUDZIŃSKI, J., DOSIAK, J. and WACLAWCZYK, J. (1973). Transcutaneous inhibition of sweat gland function by atropine. *J. appl. Physiol.* **34**, 850–852.

GILBERT, T. M. and BLATTEIS, C. M. (1977). Hypothalamic thermoregulatory pathways in the rat. *J. appl. Physiol.* **43**, 770–777.

GISOLFI, C. and ROBINSON, S. (1970). Central and peripheral stimuli regulating sweating during intermittent work in men. *J. appl. Physiol.* **29**, 761–768.

GISOLFI, C. V., WILSON, N. C., MYERS, R. D. and PHILLIPS, M. I. (1976). Exercise thermoregulation: hypothalamic perfusion of excess calcium reduces elevated colonic temperature of rats. *Brain Res.* **101**, 160–164.

GISOLFI, C. V., MORA, F. and MYERS, R. D. (1977). Diencephalic efflux of calcium ions in the monkey during exercise, thermal stress and feeding. *J. Physiol. (Lond.)* **273**, 617–630.

GLASER, E. M. (1966). "The Physiological Basis of Habituation". Oxford Univ. Press, London.

GLOTZBACH, ST. F. and HELLER, H. C. (1976). Central nervous regulation of body temperature during sleep. *Science* **194**, 537–539.

GÖBEL, D., MARTIN, H. and SIMON, E. (1977). Primary cardiac responses to stimulation of hypothalamic and spinal cord temperature sensors evaluated in anaesthetized paralyzed dogs. *J. therm. Biol.* **2**, 41–47.

GOLDSCHEIDER, A. (1898). "Gesammelte Abhandlungen". Johann Ambrosius Barth, Leipzig.

GOLENHOFEN, K. (1962). Physiologie des menschlichen Muskelkreislaufes. *Marburger Sitz.-Ber.* **83/84**, 167–254.

GOLENHOFEN, K. (1963). Zur Topographie der Muskelaktivität bei Kältebeblastung des Menschen. *Arch. phys. Ther. (Lpz.)* **15**, 435–438.

GOLENHOFEN, K. (1970). Primäre Kältewirkungen und Thermoregulation. *Z. phys. Med.* **1**, 6–21.

GOLENHOFEN, K. (1971). Haut. *In* "Lehrbuch der Physiologie", Physiologie des Kreislaufs 1 (E. Bauereisen, Ed.), pp. 347–384. Springer, Berlin, Heidelberg and New York.

GOLLWITZER-MEIER, K. (1937). Beiträge zur Wärmeregulation auf Grund von Bäderwirkungen. *Klin. Wschr.* 1937 II, 1418–1421.

GONZALEZ, R. R. (1979). Role of natural acclimatization (cold and heat) and temperature: effect on health and acceptability in a built environment. *In* "Indoor Climate" (P. O. Fanger and O. Valbjørn, Eds), pp. 737–751. Danish Building Res. Inst., Copenhagen.

GONZALEZ, R. R., PANDOLF, K. B. and GAGGE, A. P. (1973). Physiological responses and warm discomfort during heat strain. *Arch. Sci. Physiol.* **27,** A563–A571.

GONZALEZ, R. R., PANDOLF, K. B. and GAGGE, A. P. (1974). Heat acclimation and decline in sweating during humidity transients. *J. appl. Physiol.* **36,** 419–425.

GONZALEZ, R. R., NISHI, Y. and GAGGE, A. P. (1978). Mean body temperature and effective temperature as indices of human thermoregulatory response to warm environments. *In* "New Trends in Thermal Physiology" (Y. Houdas and J. D. Guieu, Eds), pp. 116–120. Masson, Paris, New York, Barcelona and Milan.

GREEN, B. G. (1977). Localization of thermal sensation: an illusion and synthetic heat. *Percept. Psychophys.* **22,** 331–337.

GREEN, B. G. (1978). Referred thermal sensations: warmth versus cold. *Sensory Processes* **2,** 220–230.

GRUNDIG, J. (1930). Zur Frage der paradoxen Warmempfindungen. *Z. Biol.* **89,** 547–554.

GUIEU, J. D. and HARDY, J. D. (1970a). Effects of preoptic and spinal cord temperature in control of thermal polypnea. *J. appl. Physiol.* **28,** 540–542.

GUIEU, J. D. and HARDY, J. D. (1970b). Effects of heating and cooling of the spinal cord on preoptic unit activity. *J. appl. Physiol.* **29,** 675–683.

GUIEU, J. D. and HARDY, J. D. (1971). Integrative activity of preoptic units I: Response to local and peripheral temperature changes. *J. Physiol. (Paris)* **63,** 253–256.

GUPTA, B. N., NIER, K. and HENSEL, H. (1979). Cold-sensitive afferents from the abdomen. *Pflügers Arch.* **380,** 203–204.

HAAHR, S. and MOGENSEN, S. (1978). Function of fever in infectious disease. *Biomed.* **28,** 305–307.

HAGEN, E., KNOCHE, H., SINCLAIR, D. and WEDDELL, G. (1953). The role of specialized nerve terminals in cutaneous sensibility. *Proc. roy. Soc. B.* **141,** 279–287.

HAHN, H. (1927). Die Reize und die Reizbedingungen des Temperatursinnes. I. Der für den Temperatursinn adäquate Reiz. *Pflügers Arch. ges. Physiol.* **215,** 133–169.

HAHN, H. (1949). "Beiträge zur Reizphysiologie". Scherer, Heidelberg.

HAHN, P. and NOVAK, M. (1975). Development of brown and white adipose tissue. *J. Lipid Res.* **16,** 79–91.

HALBERG, F. (1969). Chronobiology. *Ann. Rev. Physiol.* **31**, 675–725.

HALES, J. R. S. and IRIKI, M. (1975). Integrated changes in regional circulatory activity evoked by spinal cord and peripheral thermoreceptor stimulation. *Brain Res.* **87**, 267–279.

HALES, J. R. S., BENNET, J. W., BAIRD, J. A. and FAWCETT, A. A. (1973). Thermoregulatory effects of prostaglandins E_1, E_2, $F_{1\alpha}$, and $F_{2\alpha}$ in the sheep. *Pflügers Arch.* **339**, 125–133.

HALES, J. R. S., FAWCETT, A. A. and BENNETT, J. W. (1975). Differential influences of CNS and superficial body temperatures on the partition of cutaneous blood flow between capillaries and arteriovenous anastomoses (AVA's). *Pflügers Arch.* **361**, 105–106.

HALES, J. R. S., BENNETT, J. W. and FAWCETT, A. A. (1976). Effects of acute cold exposure on the distribution of cardiac output in the sheep. *Pflügers Arch.* **366**, 153–157.

HALES, J. R. S., BENNETT, J. W. and FAWCETT, A. A. (1977). Integrated changes in regional circulatory activity evoked by thermal stimulation of the hypothalamus. *Pflügers Arch.* **372**, 157–164.

HAMMEL, H. T. (1964). Terrestrial animals in cold: recent studies of primitive man. *In* "Handbook of Physiology" (D. B. Dill, Ed.), Sect. 4, pp. 413–434. Amer. Physiol. Soc., Washington, D.C.

HAMMEL, H. T. (1968). Regulation of internal body temperature. *Ann. Rev. Physiol.* **30**, 641–710.

HAMMEL, H. T. (1970). Concept of the adjustable set temperature. *In* "Physiological and Behavioral Temperature Regulation" (J. D. Hardy, A. P. Gagge and J. A. J. Stolwijk, Eds), pp. 676–683. Thomas, Springfield, Ill.

HAMMEL, H. T. (1972). The set-point in temperature regulation: analogy or reality. *In* "Essays on Temperature Regulation" (J. Bligh and R. E. Moore, Eds), pp. 121–137. North-Holland Publ. Comp., Amsterdam and London.

HAMMEL, H. T. and SHARP, F. (1971). Thermoregulatory salivation in the running dog in response to pre-optic heating and cooling. *J. Physiol. (Paris)* **63**, 260–263.

HAMMEL, H. T., ELSNER, R. W., LeMESSURIER, D. H., ANDERSEN, H. T. and MILAN, F. A. (1959). Thermal and metabolic responses of the Australian aborigine to moderate cold in summer. *J. appl. Physiol.* **14**, 605–615.

HAMMEL, H. T., JACKSON, D. C., STOLWIJK, J. A. J., HARDY, J. D. and STRØMME, S. B. (1963). Temperature regulation by hypothalamic proportional control with an adjustable set point. *J. appl. Physiol.* **18**, 1146–1154.

HANDWERKER, H. O., IGGO, A. and ZIMMERMANN, M. (1975). Segmental and supraspinal actions on dorsal horn neurons responding to noxious and non-noxious skin stimuli. *Pain* **1**, 147–165.

HARADA, E. and KANNO, T. (1975). Rabbit's ear in cold acclimation studies on the change in ear temperature. *J. appl. Physiol.* **38**, 389–394.

HARDY, J. D. (1961). Physiology of temperature regulation. *Physiol. Rev.* **41,** 521–606.

HARDY, J. D. (1969). Thermoregulatory responses to temperature changes in the midbrain of the rabbit. *Fed. Proc.* **28,** 713.

HARDY, J. D. (1970). Thermal comfort: skin temperature and physiological thermoregulation. *In* "Physiological and Behavioral Temperature Regulation" (J. D. Hardy, A. P. Gagge and J. A. J. Stolwijk, Eds), pp. 856–873. Thomas, Springfield, Ill.

HARDY, J. D. (1972a). Peripheral inputs to the central regulator for body temperature. *In* "Advances in Climatic Physiology" (S. Itoh, K. Ogata and H. Yoshimura, Eds), pp. 3–21. Igaku Shoin, Tokyo.

HARDY, J. D. (1972b). Models of temperature regulation. *In* "Essays on Temperature Regulation" (J. Bligh and R. E. Moore, Eds), pp. 163–186. North-Holland Publ. Comp., Amsterdam and London.

HARDY, J. D. and DuBois, E. F. (1940). Differences between men and women in their response to heat and cold. *Proc. Natl. Acad. Sci. US* **26,** 389–398.

HARDY, J. D. and OPPEL, TH. W. (1937). Studies in temperature sensation. III. *J. clin. Invest.* **16,** 533–540.

HARDY, J. D. and OPPEL, TH. W. (1938). Studies in temperature sensation. IV. *J. clin. Invest.* **17,** 771–778.

HARDY, J. D., GAGGE, A. P. and STOLWIJK, J. A. J. (Eds) 1970. "Physiological and Behavioral Temperature Regulation". Thomas, Springfield, Ill.

HARDY, J. D., STOLWIJK, J. A. J. and GAGGE, A. P. (1971). Man. *In* "Comparative Physiology of Thermoregulation" (G. C. Whittow, Ed.), Vol. 2, pp 327–380. Academic Press, New York and London.

HAYWARD, J. N. (1973). Anatomy of heat exchange. *In* "Pharmacology of Thermoregulation" (E. Schonbaum and P. Lomax, Eds), pp. 22–41. Karger, Basel.

HAYWARD, J. N. (1977). Functional and morphological aspects of hypothalamic neurons. *Physiol. Rev.* **57,** 574–658.

HAYWARD, J. N. and BAKER, A. (1969). A comparative study of the role of the cerebral arterial blood in the regulation of brain temperature in five mammals. *Brain Res.* **16,** 417–440.

HAYWARD, J. S., ECKERSON, J. D. and COLLIS, M. L. (1977). Thermoregulatory heat production in man: prediction equation based on skin and core temperatures. *J. appl. Physiol.* **42,** 377–384.

HEATH, M. E. and INGRAM, D. L. (1978). Physiological and morphological effects of rearing young pigs in hot and cold environments. *J. Physiol. (Lond.)* **284,** 8P–9P.

HEES, J. VAN (1976). Single afferent C fiber activity in the human nerve during painful and non painful skin stimulation with radiant heat. *In* "Sensory Functions of the Skin" (Y. Zotterman, Ed), pp. 503–505. Pergamon Press, Oxford.

HEES, J. VAN and GYBELS, J. M. (1972). Pain related to single afferent C fibers from human skin. *Brain Res.* **48,** 397–400.

HEISTAD, D. D., ABBOUD, F. M., MARK, A. L. and SCHMID, P. G. (1973). Interaction of thermal and baroreceptor reflexes in man. *J. appl. Physiol.* **35**, 581–586.

HELDMAIER, G. (1974a). Cold adaptation by short daily cold exposures in the young pig. *J. appl. Physiol.* **36**, 163–168.

HELDMAIER, G. (1974b). Temperature adaptation and brown adipose tissue in hairless and albino mice. *J. comp. Physiol.* **92**, 281–292.

HELDMAIER, G. (1975). The effect of short daily cold exposures on development of brown adipose tissue in mice. *J. comp. Physiol.* **98**, 161–168.

HELLBRÜGGE, T. (1960). The development of circadian rhythms in infants. *Cold Spr. Harb. Symp. quant. Biol.* **25**, 311.

HELLON, R. F. (1967). Thermal stimulation of hypothalamic neurones in unanaesthetized rabbits. *J. Physiol. (Lond.)* **193**, 381–395.

HELLON, R. F. (1969). Environmental temperature and firing rate of hypothalamic neurones. *Experientia (Basel)* **25**, 610.

HELLON, R. F. (1970a). The stimulation of hypothalamic neurones by changes in ambient temperature. *Pflügers Arch.* **321**, 56–66.

HELLON, R. F. (1970b). Hypothalamic neurons responding to changes in hypothalamic and ambient temperatures. *In* "Physiological and Behavioral Temperature Regulation" (J. D. Hardy, A. P. Gagge and J. A. J. Stolwijk, Eds), pp. 463–471. Thomas, Springfield, Ill.

HELLON, R. F. (1972a). Central thermoreceptors and thermoregulation. *In* "Handbook of Sensory Physiology" (E. Neil, Ed.), Vol. 3, pp. 161–186. Springer, Heidelberg and New York.

HELLON, R. F. (1972b). Central transmitters and thermoregulation. *In* "Essays on Temperature Regulation" (J. Bligh and R. E. Moore, Eds), pp. 71–85. North-Holland Publ. Comp., Amsterdam and London.

HELLON, R. F. (1972c). Temperature-sensitive neurons in the brain stem: their responses to brain temperature at different ambient temperatures. *Pflügers Arch.* **335**, 323–334.

HELLON, R. F. (1975). Monoamines, pyrogens and cations: their actions on central control of body temperature. *Pharmacol. Rev.* **26**, 289–321.

HELLON, R. F. and MISRA, N. K. (1973a). Neurones in the dorsal horn of the rat responding to scrotal skin temperature changes. *J. Physiol. (Lond.)* **232**, 375–388.

HELLON, R. F. and MISRA, N. K. (1973b). Neurones in the ventrobasal complex of the rat thalamus responding to scrotal skin temperature changes. *J. Physiol. (Lond.)* **232**, 389–399.

HELLON, R. F. and MITCHELL, D. (1975). Convergence in a thermal afferent pathway in the rat. *J. Physiol. (Lond.)* **248**, 359–376.

HELLON, R. F., MISRA, N. K. and PROVINS, K. A. (1973). Neurones in the somatosensory cortex of the rat responding to scrotal skin temperature changes. *J. Physiol. (Lond.)* **232**, 401–411.

HELLON, R. F., HENSEL, H. and SCHÄFER, K. (1975). Thermal receptors in the scrotum of the rat. *J. Physiol. (Lond.)* **248**, 349–357.

HENANE, R. (1972). La dépression sudorale au cours de l'hyperthermie contrôlée chez l'homme. Effets sur le débit et les électrolytes sudoraux. *J. Physiol. (Paris)* **64**, 147–163.

HENANE, R. and VALATX, J. L. (1973). Thermoregulatory changes induced during heat acclimatization by controlled hyperthermia in man. *J. Physiol. (Lond.)* **230**, 255–272.

HENANE, R., BUGUET, A., ROUSSEL, B. and BITTEL, J. (1977a). Variations in evaporation and body temperatures during sleep in man. *J. appl. Physiol.* **42**, 50–55.

HENANE, R., FLANDROIS, R. and CHARBONNIER, J. P. (1977b). Increase in sweating sensitivity by endurance conditioning in man. *J. appl. Physiol.* **43**, 822–828.

HENANE, R., BUGUET, A., BITTEL, J. and ROUSSEL, B. (1979). Thermal rhythms and sleep in man. Approach to thermal comfort during sleep. *In* "Thermal Comfort" (J. Durand and J. Raynaud, Eds), pp. 195–234. Editions INSERM, Paris.

HENSEL, H. (1950a). Die intracutane Temperaturbewegung bei Einwirkung äußerer Temperaturreize. *Pflügers Arch. ges. Physiol.* **252**, 146–164.

HENSEL, H. (1950b). Temperaturempfindung und intracutane Wärmebewegung. *Pflügers Arch. ges. Physiol.* **252**, 165–215.

HENSEL, H. (1952a). Physiologie der Thermoreception. *Ergebn. Physiol.* **47**, 166–368.

HENSEL, H. (1952b). Afferente Impulse aus den Kältereceptoren der äußeren Haut. *Pflügers Arch. ges. Physiol.* **256**, 195–211.

HENSEL, H. (1953a). Das Verhalten der Thermoreceptoren bei Temperatursprüngen. *Pflügers Arch. ges. Physiol.* **256**, 470–478.

HENSEL, H. (1953b). Das Verhalten der Thermoreceptoren bei Ischämie. *Pflügers Arch. ges. Physiol.* **257**, 371–383.

HENSEL, H. (1966). "Allgemeine Sinnesphysiologie. Hautsinne, Geschmack, Geruch". Springer, Berlin, Heidelberg and New York.

HENSEL, H. (1968). Spezifische Wärmeimpulse aus der Nasenregion der Katze. *Pflügers Arch.* **302**, 374–376.

HENSEL, H. (1969). Cutane Wärmereceptoren bei Primaten. *Pflügers Arch.* **313**, 150–152.

HENSEL, H. (1970). Temperature receptors in the skin. *In* "Physiological and Behavioral Temperature Regulation" (J. D. Hardy, A. P. Gagge and J. A. J. Stolwijk, Eds), pp. 442–462. Thomas, Springfield, Ill.

HENSEL, H. (1973a). Cutaneous thermoreceptors. *In* "Handbook of Sensory Physiology" (A. Iggo, Ed.), Vol. 2, pp. 79–110. Springer, Berlin, Heidelberg and New York.

HENSEL, H. (1973b). Neural processes in thermoregulation. *Physiol. Rev.* **53**, 948–1017.

HENSEL, H. (1973c). Temperature reception and thermal comfort. *Arch. Sci. physiol.* **27**, A359–A370.

HENSEL, H. (1974a). Thermoreception. *In* "Encyclopaedia Britannica", Vol. 18, pp. 328–332. Encyclopaedia Britannica, Inc., Chicago.

HENSEL, H. (1974b). Thermoreceptors. *Ann. Rev. Physiol.* **36**, 233–249.

HENSEL, H. (1974c). Thermische Adaptation am Menschen. *In* "Arbeitsberichte des Sonderforschungsbereichs Adaptation und Rehabilitation", Vol. 1, pp. 44–54. Marburg.

HENSEL, H. (1976a). Correlations of neural activity and thermal sensation in man. *In* "Sensory Functions of the Skin in Primates" (Y. Zotterman, Ed.), pp. 331–353. Pergamon Press, Oxford.

HENSEL, H. (1976b). Functional and structural basis of thermoreception. *In* "Progress in Brain Research" (A. Iggo and O. B. Ilyinsky, Eds), Vol. 43, pp. 105–118. Elsevier, Amsterdam, Oxford and New York.

HENSEL, M. (1977). Temperaturempfindung und Affektreaktion bei verschiedener Raumtemperatur vor und nach thermischer Langzeitadaptation. Inaug.-Diss. Marburg.

HENSEL, H. (1979a). Processing of thermal information. *In* "Thermal Comfort" (J. Durand and J. Raynaud, Eds), pp. 39–56. Editions INSERM, Paris.

HENSEL, H. (1979b). Thermoreception and human comfort. *In* "Indoor Climate" (P. O. Fanger and O. Valbjørn, Eds), pp. 425–440. Danish Building Res. Inst., Copenhagen.

HENSEL, H. and BANET, M. (1978). Thermoreceptor activity, insulative and metabolic changes in cold and warm adapted cats. *In* "New Trends in Thermal Physiology" (Y. Houdas and J. D. Guieu, Eds), pp. 53–55. Masson, Paris, New York, Barcelona and Milan.

HENSEL, H. and BOMAN, K. K. A. (1960). Afferent impulses in cutaneous sensory nerves in human subjects. *J. Neurophysiol.* **23**, 564–578.

HENSEL, H. and HUOPANIEMI, T. (1969). Static and dynamic properties of warm fibres in the infraorbital nerve. *Pflügers Arch.* **309**, 1–10.

HENSEL, H. and IGGO, A. (1971). Analysis of cutaneous warm and cold fibres in primates. *Pflügers Arch.* **329**, 1–8.

HENSEL, H. and KENSHALO, D. R. (1969). Warm receptors in the nasal region of cats. *J. Physiol. (Lond.)* **204**, 99–112.

HENSEL, H. and SCHÄFER, K. (1974). Effects of calcium on warm and cold receptors. *Pflügers Arch.* **352**, 87–90.

HENSEL, H. and SCHÄFER, K. (1979). Activity of cold receptors in cats after long-term adaptation to various temperatures. *Pflügers Arch.* **379**, R56.

HENSEL, H. and WITT, I. (1959). Spatial temperature gradient and thermoreceptor stimulation. *J. Physiol. (Lond.)* **148**, 180–189.

HENSEL, H. and ZOTTERMAN, Y. (1951a). The response of the cold receptors to constant cooling. *Acta physiol. scand.* **22**, 96–113.

HENSEL, H. and ZOTTERMAN, Y. (1951b). Quantitative Beziehungen zwischen der Entladung einzelner Kältefasern und der Temperatur. *Acta physiol. scand.* **23**, 291–319.

HENSEL, H. and ZOTTERMAN, Y. (1951c). Action potentials of cold fibres and intracutaneous temperature gradient. *J. Neurophysiol.* **14**, 377–385.

HENSEL, H. and ZOTTERMAN, Y. (1951d). The response of mechanoreceptors to thermal stimulation. *J. Physiol. (Lond.)* **115**, 16–24.

HENSEL, H. and ZOTTERMAN, Y. (1951e). The effect of menthol on the thermoreceptors. *Acta physiol. scand.* **24,** 27–34.

HENSEL, H., STRÖM, L. and ZOTTERMAN, Y. (1951). Electrophysiological measurements of depth of thermoreceptors. *J. Neurophysiol.* **14,** 423–429.

HENSEL, H., IGGO, A. and WITT, I. (1960). A quantitative study of sensitive cutaneous thermoreceptors with C afferent fibres. *J. Physiol. (Lond.)* **153,** 113–126.

HENSEL, H., BRÜCK, K. and RATHS, P. (1973). Homeothermic organisms. *In* "Temperature and Life" (H. Precht, J. Christophersen, H. Hensel and W. Larcher, Eds), pp. 505–732. Springer, Berlin, Heidelberg and New York.

HENSEL, H., ANDRES, K. H. and DÜRING, M. VON (1974). Structure and function of cold receptors. *Pflügers Arch.* **352,** 1–10.

HÉROUX, O. and PIERRE, J. ST. (1957). Effects of cold acclimation on vascularization of ears, heart, liver and muscles of white rats. *Amer. J. Physiol.* **188,** 163–168.

HEY, E. N. (1972). Thermal regulation in the newborn. *Brit. J. Hosp. Med.* 51–64.

HEY, E. N. and KATZ, G. (1969). Evaporative water loss in the new-born baby. *J. Physiol. (Lond.)* **200,** 605–619.

HEY, E. N. and KATZ, G. (1970). The range of thermal insulation in the tissues of the new-born baby. *J. Physiol. (Lond.)* **207,** 667–681.

HEY, E. N., KATZ, G. and O'CONNELL, B. (1970). The total thermal insulation of the new-born baby. *J. Physiol. (Lond.)* **207,** 683–698.

HILDEBRANDT, G. (1974). Circadian variations of thermoregulatory response in man. *In* "Chronobiology" (L. E. Scheving, F. Halberg and J. E. Pauly, Eds), pp. 234–240. Thieme, Stuttgart.

HILDEBRANDT, G. (1977). Tagesrhythmische Einflüsse auf das Adaptations-vermögen des Menschen. *In* "Arbeitsberichte des Sonderforschungs-bereiches Adaptation und Rehabilitation", pp. 157–208. Marburg.

HIRSCHSOHN, J. and MAENDL, H. (1922). Studien zur Dynamik der endove-nösen Injektion bei Anwendung von Kalzium. *Wien. Arch. inn. Med.* **4,** 379–414.

HÖFLER, W. (1968). Changes in regional distribution of sweating during acclimatization to heat. *J. appl. Physiol.* **25,** 503–506.

HONG, S.-I. and NADEL, E. R. (1979). Thermogenic control during exercise in a cold environment. *J. appl. Physiol.* **47,** 1084–1089.

HONG, S. K. (1963). Comparison of diving and nondiving women of Korea. *Fed. Proc.* **22,** 831–833.

HONG, S. K. (1973). Pattern of cold adaptation in women divers of Korea (ama). *Fed. Proc.* **32,** 1614–1622.

HORI, T. and HARADA, Y. (1976). Midbrain neuronal responses to local and spinal cord temperatures. *Amer. J. Physiol.* **231,** 1573–1578.

HORI, T. and NAKAYAMA, T. (1973). Effects of biogenic amines on central thermoresponsive neurones in the rabbit. *J. Physiol. (Lond.)* **232,** 71–85.

HORI, S., INOUYE, A., IHZUKA, H. and YAMADA, T. (1974). Study on seasonal variations of heat tolerance in young Japanese males and effects of physical training thereon. *Jap. J. Physiol.* **24,** 463–474.

HOUDAS, Y. and GUIEU, J. D. (Eds) (1978). "New Trends in Thermal Physiology". Masson, Paris, New York, Barcelona and Milan.

HOUDAS, Y., COLIN, J., TIMBAL, J., BOUTELIER, CH. and GUIEU, J. D. (1972). Skin temperatures in warm environments and the control of sweat evaporation. *J. appl. Physiol.* **33,** 99–104.

HOUDAS, Y., SAUVAGE, A., BONAVENTURE, M. and GUIEU, J. D. (1973). Modèle de la réponse évaporatoire à l'augmentation de la charge thermique. *J. Physiol. (Paris)* **66,** 137–161.

HOUDAS, Y., LECROART, J. L., LEDRU, C., CARETTE, G. and GUIEU, J. D. (1978). The thermoregulatory mechanisms considered as a follow-up system. *In* "New Trends in Thermal Physiology" (Y. Houdas and J. D. Guieu, Eds), pp. 11–19. Masson, Paris.

HULL, D. and HARDMAN, M. J. (1970). Brown adipose tissue in newborn mammals. *In* "Brown Adipose Tissue (O. Lindberg, Ed.), pp. 97–115. Amer. Elsevier Publ. Comp., New York.

IGGO, A. (1959). Cutaneous heat and cold receptors with slowly-conducting (C) afferent fibres. *Quart. J. exp. Physiol.* **44,** 362–370.

IGGO, A. (1963). An electrophysiological analysis of afferent fibres in primate skin. *Acta neuroveg. (Wien)* **24,** 225–240.

IGGO, A. (1968). Electrophysiological and histological studies of cutaneous mechanoreceptors. *In* "The Skin Senses" (D. R. Kenshalo, Ed.), pp. 84–111. Thomas, Springfield, Ill.

IGGO, A. (1969). Cutaneous thermoreceptors in primates and sub-primates. *J. Physiol. (Lond.)* **200,** 403–430.

IGGO, A. and IGGO, B. J. (1971). Impulse coding in primate cutaneous thermoreceptors in dynamic thermal conditions. *J. Physiol. (Paris)* **63,** 287–290.

IGGO, A. and RAMSEY, R. L. (1976). Thermosensory mechanisms in the spinal cord of monkeys. *In* "Sensory Functions of the Skin in Primates" (Y. Zotterman, Ed.), pp. 285–304. Pergamon Press, Oxford.

IGGO, A. and YOUNG, D. W. (1975). Cutaneous thermoreceptors and thermal nociceptors. *In* "The Somatosensory System" (H. H. Kornhuber, Ed.), pp. 5–22. Thieme, Stuttgart.

INGRAM, D. L. (1977). Adaptations to ambient temperature in growing pigs. *Pflügers Arch.* **367,** 257–264.

INGRAM, D. L. and LEGGE, K. F. (1971). The influence of deep body temperatures and skin temperatures on peripheral blood flow in the pig. *J. Physiol. (Lond.)* **215,** 693–707.

INGRAM, D. L. and LEGGE, K. F. (1972a). The influence of deep body temperatures and skin temperatures on respiratory frequency in the pig. *J. Physiol. (Lond.)* **220,** 283–296.

INGRAM, D. L. and LEGGE, K. F. (1972b). The influence of deep body and skin temperatures on thermoregulatory responses to heating of the scrotum in pigs. *J. Physiol. (Lond.)* **224,** 477–487.

INGRAM, D. L. and MOUNT, L. E. (1975). "Man and Animals in Hot Environments". Springer, Berlin, Heidelberg and New York.

INOUE, S. and MURAKAMI, N. (1976). Unit responses in the medulla oblongata of rabbit to changes in local and cutaneous temperature. *J. Physiol. (Lond.)* **259**, 339–356.

IRIKI, M. (1968). Änderung der Hautdurchblutung bei unnarkotisierten Kaninchen durch isolierte Wärmung des Rückenmarks. *Pflügers Arch. ges. Physiol.* **299**, 295–310.

IRIKI, M. and HALES, J. R. S. (1976). Spontaneous thermoregulatory oscillations in cutaneous efferent sympathetic activity. *Experientia (Basel)* **32**, 879–880.

IRIUCHIJIMA, J. and ZOTTERMAN, Y. (1960). The specificity of afferent cutaneous C fibres in mammals. *Acta physiol. scand.* **49**, 267–278.

ISSING, K. and HENSEL, H. (1979). Local thermal comfort and skin blood flow. *Pflügers Arch.* **382**, R28.

ISSING, K., BESTE, R. and HENSEL, H. (1978). Perception of local thermal stimuli during arterial occlusion. *Pflügers Arch.* **377**, R56.

ITOH, S., DOI, K. and KUROSHIMA, A. (1970). Enhanced sensitivity to noradrenaline of the Ainu. *Int. J. Biometeorol.* **14**, 195–200.

JACKSON, D. L. (1967). A hypothalamic region responsive to localized injection of pyrogens. *J. Neurophysiol.* **30**, 586–602.

JACOBSON, F. H. and SQUIRES, R. D. (1970). Thermoregulatory responses of the cat to preoptic and environmental temperatures. *In* "Physiological and Behavioral Temperature Regulation" (J. D. Hardy, A. P. Gagge and J. A. J. Stolwijk, Eds), pp. 581–596. Thomas, Springfield, Ill.

JÄRVILEHTO, T. (1973). Neural coding in the temperature sense. Human reactions to temperature changes as compared with activity in single peripheral cold fibers in the cat. *Ann. Acad. Sci. Fenn., Ser. B* **184**, 1–71.

JÄRVILEHTO, T. (1977). Neural basis of cutaneous sensations analyzed by microelectrode measurements from human peripheral nerves. *Scand. J. Psychol.* **18**, 348–359.

JAHNS, R. (1975). Types of neuronal responses in the rat thalamus to peripheral temperature changes. *Exp. Brain Res.* **23**, 157–166.

JAHNS, R. (1976). Different projections of cutaneous thermal inputs to single units of the midbrain raphe nuclei. *Brain Res.* **101**, 355–361.

JAHNS, R. (1977). Leitung und Verarbeitung thermischer Informationen im thermoafferenten System. *Habil.-Schrift*, Bochum.

JAHNS, R. and WERNER, J. (1974). Analysis of periodic components of hypothalamic spike-trains after central thermal stimulation. *Pflügers Arch.* **351**, 13–24.

JANSKÝ, L. and MUSACCHIA, X. J. (Eds) (1976). "Depressed Metabolism and Thermogenesis". Thomas, Springfield, Ill.

JASPER, H. and AJMONE-MARSAN, C. (1954). "A Stereotaxic Atlas of the Diencephalon of the Cat". National Res. Council of Canada, Ottawa.

JELL, R. M. (1973). Response of hypothalamic neurons to local temperature and to acetylcholine, noradrenaline and 5-hydroxytryptamine. *Brain Res.* **55**, 123–134.

JESSEN, C. (1967). Auslösung von Hecheln durch isolierte Wärmung des Rückenmarks beim wachen Hund. *Pflügers Arch. ges. Physiol.* **297**, 53–70.

JESSEN, C. (1971). Spinal and hypothalamic thermodetectors constituting central thermosensitivity in the conscious dog. *J. Physiol. (Paris)* **63**, 306–308.

JESSEN, C. (1976). Two-dimensional determination of thermosensitive sites within the goat's hypothalamus. *J. appl. Physiol.* **40**, 514–520.

JESSEN, C. (1977). Interaction of air temperature and core temperatures in thermoregulation of the goat. *J. Physiol. (Lond.)* **264**, 585–606.

JESSEN, C. and CLOUGH, D. P. (1973a). Evaluation of hypothalamic thermosensitivity by feedback signals. *Pflügers Arch.* **345**, 43–59.

JESSEN, C. and CLOUGH, D. P. (1973b). Assessment of spinal temperature sensitivity in conscious goats by feedback signals. *J. comp. Physiol.* **87**, 75–88.

JESSEN, C. and LUDWIG, O. (1971). Spinal cord and hypothalamus as core sensors of temperature in the conscious dog. II. Addition of signals. *Pflügers Arch.* **324**, 205–216.

JESSEN, C. and MAYER, E. TH. (1971). Spinal cord and hypothalamus as core sensors of temperature in the conscious dog. I. Equivalence of responses. *Pflügers Arch.* **324**, 189–204.

JESSEN, C. and SIMON, E. (1971). Spinal cord and hypothalamus as core sensors of temperature in the conscious dog. III. Identity of functions. *Pflügers Arch.* **324**, 217–226.

JESSEN, C., MCLEAN, J. A., CALVERT, D. T. and FINDLAY, J. D. (1972). Balanced and unbalanced temperature signals generated in spinal cord of the ox. *Amer. J. Physiol.* **222**, 1343–1347.

JOHNSON, J. M., BRENGELMANN, G. L. and ROWELL, L. B. (1976). Interactions between local and reflex influences on human forearm skin blood flow. *J. appl. Physiol.* **41**, 826–831.

JOHNSON, J. M. and PARK, M. K. (1979). Reflex control of skin blood flow by skin temperature: role of core temperature. *J. appl. Physiol.* **47**, 1188–1192.

JOHNSON, K. O., DARIAN-SMITH, I. and LAMOTTE, C. (1973). Peripheral neural determinants of temperature discrimination in man: a correlative study of responses to cooling skin. *J. Neurophysiol.* **36**, 347–370.

JOHNSON, K. O., DARIAN-SMITH, I., LAMOTTE, C., JOHNSON, B. and OLDFIELD, S. (1979). Coding of incremental changes in skin temperature by a population of warm fibers in the monkey: correlation with intensity discrimination in man. *J. Neurophysiol.* **42**, 1332–1353.

KARLBERG, P., MOORE, R. E. and OLIVER, T. K. (1965). Thermogenic and cardiovascular response of the newborn baby to noradrenaline. *Acta Paediatr. (Uppsala)* **54**, 225–238.

KASTING, N. W., VEALE, W. L. and COOPER, K. E. (1979). Development of fever in the newborn lamb. *Amer. J. Physiol.* **236**, R184–R187.

KEATINGE, W. R. (1970). Direct effects of temperature on blood vessels: their role in cold vasodilatation. In "Physiological and Behavioral Temperature Regulation" (J. D. Hardy, A. P. Gagge and J. A. J. Stolwijk, Eds), pp. 231–236. Thomas, Springfield, Ill.

KENNEDY, M. S. and BURKS, T. F. (1974). Dopamine receptors in the central thermoregulatory mechanism of the cat. Neuropharmacology **13**, 119–128.

KENSHALO, D. R. (1968). Behavioral and electrophysiological responses of cats to thermal stimuli. In "The Skin Senses" (D. R. Kenshalo, Ed.), pp. 400–422. Thomas, Springfield, Ill.

KENSHALO, D. R. (1975). The dynamic response of cold units of the cat. In "The Somatosensory System" (H. H. Kornhuber, Ed.), pp. 38–42. Thieme, Stuttgart.

KENSHALO, D. R. (1976). Correlations of temperature sensitivity in man and monkey, a first approximation. In "Sensory Functions of the Skin" (Y. Zotterman, Ed.), pp. 305–330. Pergamon Press, Oxford and New York.

KENSHALO, D. R. (1977). Age changes in touch, vibration, temperature, kinesthesis and pain sensitivity. In "Handbook of the Psychology of Aging" (J. E. Birren and K. W. Schaie, Eds), pp. 562–579. Van Nostrand Reinhold, New York.

KENSHALO, D. R. (1979a). Aging effects on cutaneous and kinesthetic sensibilities. In "Special Senses in Aging" (S. S. Han and D. H. Coons, Eds), pp. 189–217. Institute of Gerontology, Ann Arbor.

KENSHALO, D. R. (1979b). Changes in the vestibular and somesthetic systems as a function of age. In "Sensory Systems and Communication in the Elderly" (J. M. Ordy and K. Brizzee, Eds), Vol. 10, pp. 269–282. Raven Press, New York.

KENSHALO, D. R. and DUCLAUX, R. (1977). Response characteristics of cutaneous cold receptors in the monkey. J. Neurophysiol. **40**, 319–332.

KENSHALO, D. R. and GALLEGOS, E. S. (1967). Multiple temperature-sensitive spots innervated by single nerve fibers. Science **158**, 1064–1065.

KENSHALO, D. R. and HALL, E. C. (1974). Thermal thresholds of the rhesus monkey (Macaca mulatta). J. comp. physiol. Psychol. **86**, 902–910.

KENSHALO, D. R., DUNCAN, D. G. and WEYMARK, C. (1967). Thresholds for thermal stimulation of the inner thigh, footpad, and face of cats. J. comp. physiol. Psychol. **63**, 133–138.

KENSHALO, D. R., HOLMES, CH. E. and WOOD, P. B. (1968). Warm and cool thresholds as a function of rate of stimulus temperature change. Percept. Psychophys. **3**, 81–84.

KENSHALO, D. R., HENSEL, H., GRAZIADEI, P. and FRUHSTORFER, H. (1971). On the anatomy, physiology and psychophysics of the cat's temperature sensing system. In "Oral-Facial Sensory and Motor Mechanisms" (R. Dubner and Y. Kawamura, Eds), pp. 23–45. Appleton-Century-Crofts Meredith Corp., New York.

KERSLAKE, D. MCK. and COOPER, K. E. (1950). Vasodilatation in the hand in response to heating the skin elsewhere. Clin. Sci. **9**, 31–47.

KITZING, J., KUTTA, D. and BLEICHERT, A. (1968). Temperaturregulation bei langdauernder schwerer körperlicher Arbeit. *Pflügers Arch. ges. Physiol.* **301**, 241–253.

KITZING, J., BEHLING, K., BLEICHERT, A., NINOW, S., SCARPERI, M. and SCARPERI, S. (1971). Correlations between input and output of the thermoregulatory control system during rest and exercise. *Int. J. Biometeorol.* **15**, 207–211.

KLASTERSKY, J. and KASS, E. H. (1978). Is suppression of fever or hypothermia useful in experimental and clinical infections diseases? *J. infect. Dis.* **121**, 81–86.

KLUGER, M. J. (1979a). Temperature regulation, fever, and disease. *Int. Rev. Physiol. Environ. Physiol. III* **20**, 209–251.

KLUGER, M. J. (1979b). "Fever, its Biology, Evolution, and Function". Princeton Univ. Press, Princeton, N.J.

KLUGER, M. J., GONZALEZ, R. R. and STOLWIJK, J. A. J. (1973). Temperature regulation in the exercising rabbit. *Amer. J. Physiol.* **224**, 130–135.

KLUSSMANN, F. W. (1969). Der Einfluß der Temperatur auf die afferente und efferente motorische Innervation des Rückenmarks. I. Temperaturabhängigkeit der afferenten und efferenten Spontantätigkeit. *Pflügers Arch.* **305**, 295–315.

KLUSSMANN, F. W. and HENATSCH, H.-D. (1969). Der Einfluß der Temperatur auf die afferente und efferente motorische Innervation des Rückenmarks. II. Temperaturabhängigkeit der Muskelspindelfunktion. *Pflügers Arch.* **305**, 316–339.

KLUSSMANN, F. W. and PIERAU, FR.-K. (1972). Extrahypothalamic deep body thermosensitivity. *In* "Essays on Temperature Regulation" (J. Bligh and R. E. Moore, Eds), pp. 87–104. North-Holland Publ. Comp., Amsterdam and London.

KNOX, C. V., CAMPBELL, C. and LOMAX, P. (1973). Cutaneous temperature and unit activity in the hypothalamic thermoregulatory centers. *Exp. Neurol.* **40**, 717–730.

KOBAYASHI, R. M., PALKOVITS, M., JACOBOWITZ, D. M. and KOPIN, I. J. (1975). Biochemical mapping of the noradrenergic projection from the locus coeruleus. *Neurology (Minneap.)* **25**, 223–233.

KONIETZNY, F. and HENSEL, H. (1975). Warm fiber activity in human skin nerves. *Pflügers Arch.* **359**, 265–267.

KONIETZNY, F. and HENSEL, H. (1977). The dynamic response of warm units in human skin nerves. *Pflügers Arch.* **370**, 111–114.

KONIETZNY, F. and HENSEL, H. (1979). The neural basis of the sensory quality of warmth. *In* "Sensory Functions of the Skin of Humans" (D. R. Kenshalo, Ed.), pp. 241–259. Plenum Press, New York and London.

KOSAKA, M. and SIMON, E. (1968a). Kältetremor wacher, chronisch spinalisierter Kaninchen im Vergleich zum Kältezittern intakter Tiere. *Pflügers Arch.* **302**, 333–356.

KOSAKA, M. and SIMON, E. (1968b). Der zentralnervöse spinale Mechanismus des Kältezitterns. *Pflügers Arch.* **302**, 357–373.

286   THERMORECEPTION AND TEMPERATURE REGULATION

KRÖNERT, H. and PLESCHKA, K. (1976). Lingual blood flow and its hypothalamic control in the dog during panting. *Pflügers Arch.* **367**, 25–31.

KRÜGER, F. J., KUNDT, H. W., HENSEL, H. and BRÜCK, K. (1959). Das Verhalten der Hautdurchblutung bei Hypothalamuskühlung an der wachen Katze. *Pflügers Arch. ges. Physiol.* **269**, 240–247.

KUMAZAWA, T. and PERL, E. R. (1977). Primate cutaneous sensory units with unmyelinated (C) afferent fibers. *J. Neurophysiol.* **40**, 1325–1338.

KUMAZAWA, T. and PERL, E. R. (1978). Excitation of marginal and substantia gelatinosa neurons in the primate spinal cord: indications of their place in dorsal horn functional organization. *J. comp. Neurol.* **177**, 417–434.

KUMAZAWA, T., PERL, E. R., BURGESS, P. R. and WHITEHORN, D. (1975). Ascending projections from marginal zone (Lamina I) neurons of the spinal dorsal horn. *J. comp. Neurol.* **162**, 1–11.

KUNDT, H. W., BRÜCK, K. and HENSEL, H. (1957a). Das Verhalten der Hautdurchblutung bei Kühlung des vorderen Hypothalamus. *Naturwissenschaften* **44**, 496.

KUNDT, H. W., BRÜCK, K. and HENSEL, H. (1957b). Hypothalamustemperatur und Hautdurchblutung der nichtnarkotisierten Katze. *Pflügers Arch. ges. Physiol.* **264**, 97–106.

KUROSHIMA, A., DOI, K., KURAHASHI, M. and OHNO, T. (1975). In vivo lipolytic effect of glucagon in warm-adapted and cold-adapted rats. *Jap. J. Physiol.* **25**, 275–285.

LABURN, H., MITCHELL, D. and ROSENDORFF, C. (1977). Effects of prostaglandin antagonism on sodium arachidonate fever in rabbits. *J. Physiol. (Lond.)* **267**, 559–570.

LAMOTTE, R. H. and CAMPBELL, J. N. (1978). Comparison of responses of warm and nociceptive C-fiber afferents in monkey with human judgments of thermal pain. *J. Neurophysiol.* **41**, 509–528.

LANDGREN, S. (1957a). Cortical reception of cold impulses from the tongue of the cat. *Acta physiol. scand.* **40**, 202–209.

LANDGREN, S. (1957b). Convergence of tactile, thermal, and gustatory impulses on single cortical cells. *Acta physiol. scand.* **40**, 210–221.

LANDGREN, S. (1960). Thalamic neurones responding to cooling of the cat's tongue. *Acta physiol. scand.* **48**, 255–267.

LANDGREN, S. (1970). Projections from thermoreceptors into the somatosensory system of the cat's brain. In "Physiological and Behavioral Temperature Regulation" (J. D. Hardy, A. P. Gagge and J. A. J. Stolwijk, Eds), pp. 454–462. Thomas, Springfield, Ill.

LANGKILDE, G. (1979). Thermal comfort for people of high age. In "Thermal Comfort" (J. Durand and J. Raynaud, Eds), pp. 187–193. Editions INSERM, Paris.

LATIES, V. G. and WEISS, B. (1960). Behavior in the cold after acclimatization. *Science* **131**, 1891–1892.

LEBLANC, J. (1975). "Man in the Cold". Thomas, Springfield, Ill.

LEBLANC, J. and VILLEMAIRE, A. (1970). Thyroxine and noradrenaline sensitivity, cold resistance, and brown fat. *Amer. J. Physiol.* **218,** 1742–1745.

LeBlanc, J., Dulac, S., Coté, J. and Girard, B. (1975). Autonomic nervous system and adaptation in cold in man. *J. appl. Physiol.* **39,** 181–186.

LeBlanc, J., Blais, B., Barabe, B. and Coté, J. (1976). Effects of temperature and wind on facial temperature, heart rate, and sensation. *J. appl. Physiol.* **40,** 127–131.

Lebrun, J. and Marret, D. (1979). Differences in comfort sensations in spaces heated in different ways. Belgian experiments. *In* "Indoor Climate" (P. O. Fanger and O. Valbjørn, Eds), pp. 627–643. Danish Building Res. Inst., Copenhagen.

Lee, H. K. and Chai, C. Y. (1976). Temperature-sensitive neurons in the medulla oblongata of the cat. *Brain Res.* **104,** 163–165.

Lewis, P. J., Rawlins, M. D. and Reid, J. L. (1972). Acute thermoregulatory and cardiovascular effects of 6-hydroxydopamine administered centrally in rabbits and cats. *Brit. J. Pharmacol.* **46,** 559P.

Libert, J. P., Candas, V. and Vogt, J. J. (1978). Sweating response in man during transient rises of air temperature. *J. appl. Physiol.* **44,** 284–290.

Libert, J. P., Candas, V. and Vogt, J. J. (1979). Effect of rate of change in skin temperature on local sweating rate. *J. appl. Physiol.* **47,** 306–311.

Light, A. R. and Perl, E. R. (1977). Differential termination of large-diameter and small-diameter primary afferent fibers in the spinal dorsal gray matter as indicated by labeling with horseradish perioxidase. *Neuroscience Letters* **6,** 59–63.

Light, A. R. and Perl, E. R. (1979a). Reexamination of the dorsal root projection to the spinal dorsal horn including observations on the differential termination of coarse and fine fibers. *J. comp. Neurol.* **186,** 117–132.

Light, A. R. and Perl, E. R. (1979b). Spinal termination of functionally identified primary afferent neurons with slowly conducting myelinated fibers. *J. comp. Neurol.* **186,** 133–150.

Light, A. R., Trevino, D. L. and Perl, E. R. (1979). Morphological features of functionally defined neurons in the marginal zone and substantia gelatinosa of the spinal dorsal horn. *J. comp. Neurol.* **186,** 151–172.

Liljestrand, G. and Magnus, R. (1922). Die Wirkungen des Kohlensäurebades beim Gesunden nebst Bemerkungen über den Einfluß des Hochgebirges. *Pflügers Arch. ges. Physiol.* **193,** 527–554.

Lin, M. T., Yin, T. H. and Chai, C. Y. (1972). Effects of heating and cooling of spinal cord on CV and respiratory responses and food and water intake. *Amer. J. Physiol.* **223,** 626–631.

Lind, A. R. (1977). Human tolerance to hot climates. *In* "Handbook of Physiology" (D. H. K. Lee, Ed.), Sect. 9, pp. 93–109. Amer. Physiol. Soc., Bethesda, Md.

Lindvall, O. and Björklund, A. (1974). The organization of the ascending catecholamine neuron systems in the rat brain. *Acta physiol. scand.* Suppl. 412.

LIPTON, J. M. (1968). Effect of preoptic lesions on heat escape responding and colonic temperature in the rat. *Physiol. Behav.* **3**, 165–169.

LIPTON, J. M. (1971). Behavioral temperature regulation in the rat: effects of thermal stimulation of the medulla. *J. Physiol. (Paris)* **63**, 325–328.

LIPTON, J. M. (1973). Thermosensitivity of medulla oblongata in control of body temperature. *Amer. J. Physiol.* **224**, 890–897.

LIPTON, J. M. (1980). "Fever". Raven Press, New York.

LIPTON, J. M. and FOSSLER, D. E. (1974). Fever produced in the squirrel monkey by intravenous and intracerebral endotoxin. *Amer. J. Physiol.* **226**, 1022–1027.

LIPTON, J. M. and TICKNOR, C. B. (1979). Influence of sex and age on febrile responses to peripheral and central administration of pyrogens in the rabbit. *J. Physiol. (Lond.)* **295**, 263–272.

LIPTON, J. M. and TRZCINKA, G. P. (1976). Persistence of febrile response to pyrogens after PO/AH lesions in squirrel monkeys. *Amer. J. Physiol.* **231**, 1638–1648.

LIPTON, J. M., DWYER, P. E. and FOSSLER, D. E. (1974). Effects of brainstem lesions on temperature regulation in hot and cold environments. *Amer. J. Physiol.* **226**, 1356–1365.

LIPTON, J. M., DINARELLO, C. A. and KENNEDY, J. I. (1979). Fever produced in squirrel monkeys by human leucocytic pyrogen. *Proc. Soc. exp. Biol. Med.* **160**, 426–428.

LIVINGSTONE, S. D. (1976). Changes in cold-induced vasodilatation during Arctic exercises. *J. appl. Physiol.* **40**, 455–457.

LLOYD, D. P. C., CHANG, H.-T. (1948). Afferent fibers in muscle nerves. *J. Neurophysiol.* **11**, 199–207.

LÖFSTEDT, B. E. (1979). The relation between skin temperature, blood flow and temperature sensation of the calf and foot, and local temperature exposure at different states of whole body heat balance. *In* "Indoor Climate" (P. O. Fanger and O. Valbjørn, Eds), pp. 581–589. Danish Building Res. Inst., Copenhagen.

LWOFF, A. (1969). Death and transfiguration of a problem. *Bact. Rev.* **33**, 390–403.

MacINTYRE, B. A., BULLARD, R. W., BANERJEE, M. and ELIZONDO, R. (1968). Mechanism of enhancement of eccrine sweating by localized heating. *J. appl. Physiol.* **25**, 225–260.

MAEDA, T., PIN, C., SALVERT, D., LIGIER, M. and JOUVET, M. (1973). Les neurones contenant des catécholamines du tegmentum pontique et leurs voies de projection chez le chat. *Brain Res.* **57**, 119–152.

MAEKUBO, H., MORIYA, K. and HIROSHIGE, T. (1977). Role of ketone bodies in nonshivering thermogenesis in cold-acclimated rats. *J. appl. Physiol.* **42**, 159–165.

MAGILTON, J. H. and SWIFT, C. S. (1969). Responses of veins draining the nose to alar-fold temperature changes in the dog. *J. appl. Physiol.* **27**, 18–20.

MARCUS, P. (1972). Heat acclimatization by exercise-induced elevation of body temperature. *Amer. J. Physiol.* **33**, 283–288.

MARÉCHAUX, E. W. and SCHÄFER, K. E. (1949). Über Temperaturempfindungen bei Einwirkung von Temperaturreizen verschiedener Steilheit auf den ganzen Körper. *Pflügers Arch. ges. Physiol.* **251**, 765–784.

MARKS, L. E. and GONZALEZ, R. R. (1974). Skin temperature modifies the pleasantness of thermal stimuli. *Nature (Lond.)* **247**, 473–475.

MARKS, L. E. and STEVENS, J. C. (1968). Perceived warmth and skin temperature as functions of the duration and level of thermal irradiation. *Percept. Psychophys.* **4**, 220–228.

MARKS, L. E. and STEVENS, J. C. (1973). Spatial summation of warmth: influence of duration and configuration of the stimulus. *Amer. J. Psychol.* **86**, 251–267.

MARON, M. B., WAGNER, J. A. and HORVATH, S. M. (1977). Thermoregulatory responses during competitive marathon running. *J. appl. Physiol.* **42**, 900–914.

MARTIN, B. J., MORGAN, E. J., ZWILLICH, C. W. and WEIL, J. V. (1979). Influence of exercise hyperthermia on exercise breathing pattern. *J. appl. Physiol.* **47**, 1039–1042.

MARTIN, H. F. and MANNING, J. W. (1971). Thalamic 'warming' and 'cooling' units responding to cutaneous stimulation. *Brain Res.* **27**, 377–381.

MARUHASHI, I., MIZUGUCHI, K. and TASAKI, I. (1952). Action currents in single afferent nerve fibres elicited by stimulation of the skin of the toad and the cat. *J. Physiol. (Lond.)* **117**, 129–151.

McCAFFREY, T. V., McCOOK, R. D. and WURSTER, R. D. (1975a). Effect of head skin temperature on tympanic and oral temperature in man. *J. appl. Physiol.* **39**, 114–118.

McCAFFREY, T. V., WURSTER, R. D. and GEIS, G. S. (1975b). Reflex control of sweating by skin temperature in man. *Biometeorology* **6**, 21. (Suppl. to Vol. 19 of the *Int. J. Biometeorol.*)

McCAFFREY, T. V., WURSTER, R. D., JACOBS, H. K., EULER, D. E. and GEIS, G. S. (1979). Role of skin temperature in the control of sweating. *J. appl. Physiol.* **47**, 591–597.

McCOOK, R. D., RANDALL, W. C., HASSLER, C. R., MIHALDZIC, N. and WURSTER, R. D. (1970). The role of cutaneous thermal receptors in sudomotor control. In "Physiological and Behavioral Temperature Regulation" (J. D. Hardy, A. P. Gagge and J. A. J. Stolwijk, Eds), pp. 627–633. Thomas, Springfield, Ill.

McINTYRE, D. A. (1976). Thermal sensation. A comparison of rating scales and cross modality matching. *Int. J. Biometeorol.* **20**, 295–303.

McINTYRE, D. A. (1979). The effect of air movement on thermal comfort and sensation. In "Indoor Climate" (P. O. Fanger and O. Valbjørn, Eds), pp. 541–560. Danish Building Res. Inst., Copenhagen.

McLEAN, J. A., HALES, J. R. S., JESSEN, C. and CALVERT, D. T. (1970). Influences of spinal cord temperature on heat exchange of the ox. *Proc. Aust. Physiol. Pharmacol. Soc.* **1**, 32–33.

MELCHIOR, H. and HILDEBRANDT, G. (1967). Die Hautdurchblutung verschiedener Körperregionen bei Arbeit. *Int. Z. angew. Physiol.* **24**, 68–80,

MERCER, J. B. and JESSEN, C. (1978a). Effects of total body core cooling on heat production of conscious goats. *Pflügers Arch.* **373**, 259–267.

MERCER, J. B. and JESSEN, C. (1978b). Central thermosensitivity in conscious goats: hypothalamus and spinal cord versus residual inner body. *Pflügers Arch.* **374**, 179–186.

MERCER, J. B. and JESSEN, C. (1979). Control of respiratory evaporative heat loss in exercising goats. *J. appl. Physiol.* **46**, 978–983.

MERCER, J. B., JESSEN, C. and PIERAU, FR.-K. (1978). Thermal stimulation of neurons in the rostral brain stem of conscious goats. *J. therm. Biol.* **3**, 5–10.

MERKER, G., ZEISBERGER, E., BERGHEIM-HACKMANN, E. and BRÜCK, K. (1977). Horseradish peroxidase (HRP) tracing of ascending projections from lower brain stem to thermointegrative structures of the hypothalamus. *Pflügers Arch.* **368**, R29.

MERKLIN, R. J. (1974). Growth and distribution of human fetal brown fat. *Anat. Rec.* **178**, 637–646.

METCALF, G. and MYERS, R. D. (1976). A comparison between the hypothermia induced by intraventricular injections of thyrotropin releasing hormone, noradrenaline or calcium ions in unanaesthetized cats. *Brit. J. Pharmacol.* **58**, 489–495.

METCALF, G. and MYERS, R. D. (1978). Precise location within the preoptic area where noradrenaline produces hypothermia. *Europ. J. Pharmacol.* **51**, 47–53.

MEURER, K.-A., JESSEN, C. and IRIKI, M. (1967). Kältezittern während isolierter Kühlung des Rückenmarks nach Durchschneidung der Hinterwurzeln. *Pflügers Arch. ges. Physiol.* **293**, 236–255.

MILTON, A. S. and WENDLANDT, S. (1971). Effects on body temperature of prostaglandins of the A, E and F series on injection into the third ventricle of unanaesthetized cats and rabbits. *J. Physiol. (Lond.)* **218**, 325–336.

MINUT-SOROKHTINA, O. P. (1968). The nature of rhythmic activity of cold receptors. *Sechenov Physiol. J. USSR* **54**, 413–420.

MINUT-SOROKHTINA, O. P. (1972). "Physiology of Thermoreception" (Russ.). Medicine, Moscow.

MINUT-SOROKHTINA, O. P. (1977). Enteroceptive and exteroceptive signalization in the thermoregulatory system. *In* "Modern Tendencies in Neurophysiology" (Russ.), pp. 179–188. Science, Leningrad.

MINUT-SOROKHTINA, O. P. (1978). Neurophysiology of thermoreception. *In* "Sensory Systems: Thermoreception" (Russ.), pp. 82–101. Science, Leningrad.

MITCHELL, D., ATKINS, A. R. and WYNDHAM, C. H. (1972). Mathematical and physical models of Thermoregulation. *In* "Essays on Temperature Regulation" (J. Bligh and R. E. Moore, Eds), pp. 37–54. North-Holland Publ. Comp., Amsterdam and London.

MITCHELL, D., SNELLEN, J. W. and ATKINS, A. R. (1970). Thermoregulation during fever: change of set-point or change of gain. *Pflügers Arch.* **321**, 293–302.

MOLINARI, H. H. and KENSHALO, D. R. (1977). Effect of cooling rate on the dynamic response of cat cold units. *Exp. Neurol.* **55**, 546–555.

MOLINARI, H. H., RÓZSA, A. J. and KENSHALO, R. (1976). Rhesus monkey (Macaca mulatta) cool sensitivity measured by a signal detection method. *Percept. Psychophys.* **19**, 246–251.

MOLINARI, H. H., GREENSPAN, J. D. and KENSHALO, D. R. (1977). The effects of rate of temperature change and adapting temperature on thermal sensitivity. *Sensory Processes* **1**, 354–362.

MORIMOTO, T., SHIRAKI, K. and ASAYAMA, M. (1974). Seasonal difference in responses of body fluids to heat stress. *Jap. J. Physiol.* **24**, 249–262.

MORIYA, K., MAEKUBO, H. and ITOH, S. (1974). Turnover rate of plasma free fatty acids in cold-acclimated rats. *Jap. J. Physiol.* **24**, 419–431.

MOUNT, L. E. (1968). "The Climatic Physiology of the Pig". Edward Arnold, London.

MOWER, G. D. (1976). Perceived intensity of peripheral thermal stimuli is independent of internal body temperature. *J. comp. physiol. Psychol.* **90**, 1152–1155.

MÜLLER, J. (1840). "Handbuch der Physiologie des Menschen", Vol. 1. Hölscher, Koblenz.

MURAKAMI, N. (1973). Effects of iontophoretic application of 5-hydroxytryptamine, noradrenaline and acetylcholine upon hypothalamic temperature-sensitive neurones in rats. *Jap. J. Physiol.* **23**, 435–446.

MURAKAMI, N., UCHIMURA, H. and SAKATA, Y. (1979). Comparison of thermosensitivity in the preoptic/anterior hypothalamic area and the midbrain of the rabbit. *Pflügers Arch.* **379**, 113–116.

MURGATROYD, D. and HARDY, J. D. (1970). Central and peripheral temperatures in behavioral thermoregulation of the rat. *In* "Physiological and Behavioral Temperature Regulation" (J. D. Hardy, A. P. Gagge and J. A. J. Stolwijk, Eds), pp. 874–891. Thomas, Springfield, Ill.

MYERS, R. D. (1976a). Diencephalic efflux of $^{22}Na^+$ and $^{45}Ca^{2+}$ ions in the febrile cat: effect of an antipyretic. *Brain Res.* **103**, 412–417.

MYERS, R. D. (1976b). Chemical control of body temperature by the hypothalamus: a model and some mysteries. *Proc. Austral. Physiol. Pharmacol. Soc.* **7**, 15–32.

MYERS, R. D. (1977). New aspects of the role of hypothalamic calcium ions, 5-HT and PGE during normal thermoregulation and pyrogen fever. *In* "Drugs, Biogenic Amines and Body Temperature", pp. 51–53. Karger, Basel.

MYERS, R. D. (1978). Hypothalamic actions of 5-hydroxytryptamine neurotoxins: feeding, drinking, and body temperature. *Ann. N.Y. Acad. Sci.* **305**, 556–575.

MYERS, R. D. and BELESLIN, D. B. (1971). Changes in serotonin release in hypothalamus during cooling or warming of the monkey. *Amer. J. Physiol.* **220**, 1746–1754.

MYERS, R. D. and CHINN, C. (1973). Evoked release of hypothalamic norepinephrine during thermoregulation in the cat. *Amer. J. Physiol.* **224,** 230–236.

MYERS, R. D. and SHARPE, L. G. (1968). Temperature in the monkey: transmitter factors released from the brain during thermoregulation. *Science* **161,** 572–573.

MYERS, R. D. and VEALE, W. L. (1970). Body temperature: possible ionic mechanism in the hypothalamus controlling the set point. *Science* **170,** 95–97.

MYERS, R. D. and VEALE, W. L. (1971). The role of sodium and calcium ions in the hypothalamus in the control of body temperature of the unanaesthetized cat. *J. Physiol. (Lond.)* **212,** 411–430.

MYERS, R. D. and WALLER, M. B. (1973). Differential release of acetylcholine from the hypothalamus and mesencephalon of the monkey during thermoregulation. *J. Physiol. (Lond.)* **230,** 273–294.

MYERS, R. D. and WALLER, M. B. (1975). 5-HT- and norepinephrine-induced release of ACh from the thalamus and mesencephalon of the monkey during thermoregulation. *Brain Res.* **84,** 47–61.

MYERS, R. D. and WALLER, M. B. (1976). Is prostaglandin fever mediated by the presynaptic release of hypothalamic 5-HT or norepinephrine? *Brain Res. Bull.* **1,** 47–56.

MYERS, R. D. and YAKSH, T. L. (1969). Control of body temperature in the unanaesthetized monkey by cholinergic and aminergic systems in the hypothalamus. *J. Physiol. (Lond.)* **202,** 483–500.

MYERS, R. D. and YAKSH, T. L. (1971). Thermoregulation around a new 'set-point' established in the monkey by altering the ratio of sodium to calcium ions within the hypothalamus. *J. Physiol. (Lond.)* **218,** 609–633.

MYERS, R. D., RUDY, T. A. and YAKSH, T. L. (1971a). Fever in the monkey produced by the direct action of pyrogen on the hypothalamus. *Experientia (Basel)* **27,** 160–161.

MYERS, R. D., RUDY, T. A. and YAKSH, T. L. (1971b). Effect in the rhesus monkey of salicylate on centrally-induced endotoxin fevers. *Neuropharmacology* **10,** 775–778.

MYERS, R. D., VEALE, W. L. and YAKSH, T. L. (1971c). Changes in body temperature of the unanaesthetized monkey produced by sodium and calcium ions perfused through the cerebral ventricles. *J. Physiol. (Lond.)* **217,** 381–392.

MYERS, R. D., RUDY, T. A. and YAKSH, T. L. (1974). Fever produced by endotoxin injected into the hypothalamus of the monkey and its antagonism by salicylate. *J. Physiol. (Lond.)* **243,** 167–193.

MYERS, R. D., SIMPSON, C. W., HIGGINS, D., NATTERMANN, R. A., RICE, J. C., REDGRAVE, P. and METCALF, G. (1976). Hypothalamic Na^+ and Ca^{++} ions and temperature set-point: New mechanisms of action of a central or peripheral thermal challenge and intrahypothalamic 5-HT, NE, PGE_1 and pyrogen. *Brain Res. Bull.* **1,** 301–327.

NADEL, E. R. (1979). Sensitivity to central and peripheral stimulations in humans. *In* "Thermal Comfort" (J. Durand and J. Raynaud, Eds), pp. 57–66. Editions INSERM, Paris.

NADEL, E. R. and STOLWIJK, J. A. J. (1971). Physiologic control of rate of local sweat secretion in man. *J. Physiol. (Paris)* **63**, 353–355.

NADEL, E. R. and STOLWIJK, J. A. J. (1973). Effect of skin wettedness on sweat gland response. *J. appl. Physiol.* **35**, 689–694.

NADEL, E. R., HORVATH, ST. M., DAWSON, CH. A. and TUCKER, A. (1970). Sensitivity to central and peripheral thermal stimulation in man. *J. appl. Physiol.* **29**, 603–609.

NADEL, E. R., BULLARD, R. W. and STOLWIJK, J. A. J. (1971a). Importance of skin temperature in the regulation of sweating. *J. appl. Physiol.* **31**, 80–87.

NADEL, E. R., MITCHELL, J. W. and STOLWIJK, J. A. J. (1971b). Control of local and total sweating during exercise transients. *Int. J. Biometeorol.* **15**, 201–206.

NADEL, E. R., MITCHELL, J. W. and STOLWIJK, J. A. J. (1973). Differential thermal sensitivity in the human skin. *Pflügers Arch.* **340**, 71–76.

NADEL, E. R., PANDOLF, K. B., ROBERTS, M. F. and STOLWIJK, J. A. J. (1974). Mechanisms of thermal acclimation to exercise and heat. *J. appl. Physiol.* **37**, 515–520.

NADEL, E. R., CAFARELLI, E., ROBERTS, M. F. and WENGER, C. B. (1979). Circulatory regulation during exercise in different ambient temperatures. *J. appl. Physiol.* **46**, 430–437.

NAKAYAMA, T. and HARDY, J. D. (1969). Unit responses in the rabbit's brain stem to changes in brain and cutaneous temperature. *J. appl. Physiol.* **27**, 848–857.

NAKAYAMA, T. and HORI, T. (1973). Effects of anaesthetics and pyrogen on thermally sensitive neurons in the brainstem. *J. appl. Physiol.* **34**, 351–355.

NAKAYAMA, T., EISENMAN, J. S. and HARDY, J. D. (1961). Single unit activity of anterior hypothalamus during local heating. *Science* **134**, 560–561.

NAKAYAMA, T., HAMMEL, H. T., HARDY, J. D. and EISENMAN, J. S. (1963). Thermal stimulation of electrical activity of single units in the preoptic region. *Amer. J. Physiol.* **204**, 1122–1126.

NAKAYAMA, T., ISHIKAWA, Y. and TSURUTANI, T. (1979). Projection of scrotal thermal afferents to the preoptic and hypothalamic neurons in rats. *Pflügers Arch.* **380**, 59–64.

NAPOLITANO, L. (1965). The fine structure of adipose tissues. *In* "Handbook of Physiology" (A. E. Renold and G. F. Cahill, Jr, Eds), Sect. 5, pp. 109–123. Amer. Physiol. Soc., Washington, D.C.

NAUTA, W. J. H. and KUYPERS, H. G. J. M. (1958). Some ascending pathways in the brain stem reticular formation of the cat. *In* "Reticular Formation of the Brain" (H. H. Jaspers, L. D. Proctor, R. S. Knighton, W. C. Noshay and R. T. Costello, Eds), pp. 3–30. Little, Brown, Boston, Mass.

NIELSEN, B. (1969). Thermoregulation in rest and exercise. *Acta physiol. scand.* Suppl. 323.

NIELSEN, B. (1976). Metabolic reactions to changes in core and skin temperature in man. *Acta physiol. scand.* **97**, 129–138.

NIELSEN, B., NIELSEN, S. L. and PETERSEN, F. B. (1972). Thermoregulation during positive and negative work at different environmental temperatures. *Acta physiol. scand.* **85**, 249–257.

NIELSEN, B., SCHWARTZ, P. and ALHEDE, J. (1973). Is fever in man reflected in changes in cerebrospinal fluid concentrations of sodium and calcium ions? *Scand. J. clin. Lab. Invest.* **32**, 309–310.

NINOMIYA, I. and FUJITA, S. (1976). Reflex effects of thermal stimulation on sympathetic nerve activity to skin and kidney. *Amer. J. Physiol.* **230**, 271–278.

NOBIN, A. and BJÖRKLUND, A. (1973). Topography of the monoamine neuron system in the human brain as revealed in fetuses. *Acta physiol. scand.* Suppl. 338.

NORMELL, L. A. (1974). Distribution of impaired cutaneous vasomotor and sudomotor function in paraplegic man. *Scand. J. clin. Lab. Invest.* **133**, Suppl. 138, 25–41.

NUTIK, ST. L. (1971). Effect of temperature change of the preoptic region and skin on posterior hypothalamic neurons. *J. Physiol. (Paris)* **63**, 368–370.

NUTIK, ST. L. (1973a). Posterior hypothalamic neurons responsive to preoptic region thermal stimulation. *J. Neurophysiol.* **36**, 238–249.

NUTIK, ST. L. (1973b). Convergence of cutaneous and preoptic region thermal afferents on posterior hypothalamic neurons. *J. Neurophysiol.* **36**, 250–257.

OGATA, K. (1971). Physiological approach to the adaptability to cold and heat. *Kumamoto Bull. inst. const. Med.* **21**, Suppl.

OGAWA, T. (1970). Local effect of skin temperature on threshold concentration of sudorific agents. *J. appl. Physiol.* **28**, 18–22.

OGAWA, T. (1974). Generalized response in sweat rate to periodic cutaneous heating—with special reference to its relationship to heat tolerance. *Jap. J. Physiol.* **24**, 475–489.

OLESEN, B. W. (1977). Thermal comfort requirements for floors occupied by people with bare feet. *ASHRAE Trans.* **83** (2) No. 2451.

OLESEN, B. W. and THORSHAUGE, J. (1979). Differences in comfort sensations in spaces heated by different methods: Danish experiments. *In* "Indoor Climate" (P. O. Fanger and O. Valbjørn, Eds), pp. 645–676. Danish Building Res. Inst., Copenhagen.

OLESEN, S., FANGER, P. O., JENSEN, P. B. and NIELSEN, O. J. (1973). Comfort limits for man exposed to asymmetric thermal radiation. *In* "Proceedings of the CIB Commission W 45 Symposium: Thermal Comfort and Moderate Heat Stress". HMSO, London.

OLESEN, B. W., SCHØLER, M. and FANGER, P. O. (1979). Discomfort caused by vertical air temperature differences. *In* "Indoor Climate" (P. O. Fanger and O. Valbjørn, Eds), pp. 561–579. Danish Building Res. Inst., Copenhagen.

OSTERGAARD, J., FANGER, P. O. and OLESEN, S. (1974). The effect on man's comfort of a uniform air flow from different directions. *ASHRAE Trans.* **80**(2), 142–157.

PALMER, J. D. (1976). "An Introduction to Biological Rhythms". Academic Press, New York, San Francisco and London.

PARMEGGIANI, P. L. (1977). Interaction between sleep and thermoregulation. *Waking and Sleeping* **1**, 123–132.

PARMEGGIANI, P. L. and FRANZINI, C. (1971). Changes in the activity of hypothalamic units during sleep at different environmental temperatures. *Brain Res.* **29**, 347–350.

PARMEGGIANI, P. L. and FRANZINI, C. (1973). On the functional significance of subcortical single unit activity during sleep. *Electroenceph. clin. Neurophysiol.* **34**, 495–508.

PARMEGGIANI, P. L. and RABINI, C. (1967). Shivering and panting during sleep. *Brain Res.* **6**, 789–791.

PARMEGGIANI, P. L. and SABATTINI, L. (1972). Electromyographic aspects of postural, respiratory and thermoregulatory mechanisms in sleeping cats. *Electroenceph. clin. Neurophysiol.* **33**, 1–13.

PARMEGGIANI, P. L., FRANZINI, C., LENZI, P. and ZAMBONI, G. (1973). Threshold of respiratory responses to preoptic heating during sleep in freely moving cats. *Brain Res.* **52**, 189–201.

PARMEGGIANI, P. L., CIANCI, T., CALASSO, M. and ZAMBONI, G. (1974). Ear skin temperature changes during fast wave sleep in cats. *Experientia (Basel)* **30**(1) 682.

PARMEGGIANI, P. L., FRANZINI, C. and LENZI, P. (1976). Respiratory frequency as a function of preoptic temperature during sleep. *Brain Res.* **111**, 253–260.

PARMEGGIANI, P. L., ZAMBONI, G., CIANCI, T. and CALASSO, M. (1977). Absence of thermoregulatory vasomotor responses during fast wave sleep in cats. *Electroenceph. clin. Neurophysiol.* **42**, 372–380.

PEIRCE, C. S. (1960). "Collected Papers" (C. Hartshorne and P. Weiss, Eds), Vol. 5. Belknap Press of Harvard Univ. Press, Cambridge, Mass.

PERL, E. R. (1968). Myelinated afferent fibres innervating the primate skin and their response to noxious stimuli. *J. Physiol. (Lond.)* **197**, 593–615.

PERL, E. R. (1976). Sensitization of nociceptors and its relation to sensation. *Pain Res. Ther.* **1**, 17–28.

PIERAU, F.-K. and KLUSSMANN, F. W. (1971). Spinal excitation and inhibition during local spinal cooling and warming. *J. Physiol. (Paris)* **63**, 380–382.

PIERAU, F.-K., KLEE, M. R., FABER, D. S. and KLUSSMANN, F. W. (1971). Mechanism of cellular thermoreception in mammals. *Int. J. Biometeorol.* **15**, 134–140.

PIERAU, F.-K., TORREY, P. and CARPENTER, D. (1975). Effect of ouabain and potassium-free solution on mammalian thermosensitive afferents in vitro. *Pflügers Arch.* **359**, 349–356.

PIERAU, F.-K., ULLRICH, J. and WURSTER, R. D. (1977). Effect of Ca^{++} and EDTA on the bursting pattern of lingual cold receptors in cats. *Proc. int. Un. physiol. Sci.* **13**, 597.

PIN, C., JONES, B. and JOUVET, M. (1968). Topographie des neurones monoaminergique du tronc cérébral du chat: étude par histofluorescence. *C.R. Soc. Biol. (Paris)* **162**, 2136–2141.

PITTENDRIGH, C. S. (1974). Circadian oscillations in cells and the circadian organization of multicellular systems. *In* "The Neurosciences: Third Study Program" (F. O. Schmitt and F. G. Worden, Eds), pp. 437–458. MIT Press, Cambridge, Mass.

PITTMAN, Q. J., COOPER, K. E., VEALE, W. L. and PETTEN, G. R. VAN (1974). Observations on the development of the febrile response to pyrogens in sheep. *Clin. Sci. molec. Med.* **46**, 591–602.

PITTMAN, Q. J., VEALE, W. L. and COOPER, K. E. (1975). Temperature responses of lambs after centrally injected prostaglandins and pyrogens. *Amer. J. Physiol.* **228**, 1034–1038.

PITTMAN, Q. J., VEALE, W. L. and COOPER, K. E. (1977). Effect of prostaglandin, pyrogen and noradrenaline, injected into the hypothalamus, on thermoregulation in newborn lambs. *Brain Res.* **128**, 473–483.

PLESCHKA, K., KÜHN, P. and NAGAI, M. (1979). Differential vasomotor adjustments in the evaporative tissues of the tongue and nose in the dog under heat load. *Pflügers Arch.* **382**, 255–262.

POLETTI, C. E., KINNARD, M. A. and MACLEAN, P. D. (1973). Hippocampal influence on unit activity of hypothalamus, preoptic region, and basal forebrain in awake, sitting squirrel monkeys. *J. Neurophysiol.* **36**, 308–324.

POMPEIANO, O. (1973). Reticular formation. *In* "Handbook of Sensory Physiology" (A. Iggo, Ed.), Vol. 2, pp. 382–488. Springer, Berlin, Heidelberg and New York.

POULOS, D. A. (1971). Trigeminal temperature mechanisms. *In* "Oral-Facial Sensory and Motor Mechanisms" (R. Dubner and Y. Kawamura, Eds), pp. 47–72. Appleton-Century-Crofts Meredith Corp., New York.

POULOS, D. A. (1975). Central processing of peripheral temperature information. *In* "The Somatosensory System" (H. H. Kornhuber, Ed.), pp. 78–93. Thieme, Stuttgart.

POULOS, D. A. and BENJAMIN, R. M. (1968). Response of thalamic neurons to thermal stimulation of the tongue. *J. Neurophysiol.* **31**, 28–43.

POULOS, D. A. and MOLT, J. T. (1976). Response of central trigeminal neurons to cutaneous thermal stimulation. *In* "Sensory Functions of the Skin in Primates" (Y. Zotterman, Ed.), pp. 263–283. Pergamon Press, Oxford.

POULOS, D. A., STROMINGER, N., PELLETIER, V. and GABRIELE, T. (1970). Response of medullary neurons to thermal stimulation of the tongue. *Fed. Proc.* **29**, 392.

PROPPE, D. W., BRENGELMANN, G. L. and ROWELL, L. B. (1976). Control of baboon limb blood flow and heart rate—role of skin vs. core temperature. *Amer. J. Physiol.* **231**, 1457–1464.

PUSCHMANN, ST. and JESSEN, C. (1978). Anterior and posterior hypothalamus: effects of independent temperature displacements on heat production in conscious goats. *Pflügers Arch.* **373**, 59–68.

RAWSON, R. O. and QUICK, K. P. (1970). Evidence of deep-body thermoreceptor response to intraabdominal heating of the ewe. *J. appl. Physiol.* **28**, 813–820.

RAWSON, R. O. and QUICK, K. P. (1971a). Thermoregulatory responses to temperature signals from the abdominal viscera of sheep. *J. Physiol. (Paris)* **63**, 399–402.

RAWSON, R. O. and QUICK, K. P. (1971b). Unilateral splanchnotomy: its effect on the response to intra-abdominal heating in the ewe. *Pflügers Arch.* **330**, 362–365.

RAWSON, R. O. and QUICK, K. P. (1972). Localization of intra-abdominal thermoreceptors in the ewe. *J. Physiol. (Lond.)* **222**, 665–677.

RAWSON, R. O. and QUICK, K. P. (1976). Intra-abdominal thermoreceptors and the regulation of body temperature in sheep. *Israel J. Med. Sci.* **12**, 1040–1043.

RAYNAUD, J., MARTINEAUD, J. P., BHATNAGAR, O. P., VIELLEFOND, H. and DURAND, J. (1976). Body temperatures during rest and exercise in residents and sojourners in hot climate. *Int. J. Biometeorol.* **20**, 309–317.

REAVES, T. A. JR. (1977). Gain of thermosensitive neurons in the preoptic area of the rabbit, Oryctolagus cuniculus. *J. therm. Biol.* **2**, 31–33.

REAVES, T. A. and HEATH, J. E. (1975). Interval coding of temperature by CNS neurones in thermoregulation. *Nature (Lond.)* **257**, 688–690.

REENPÄÄ, Y. (1962). "Allgemeine Sinnesphysiologie". Klostermann, Frankfurt.

REIN, F. H. (1925a). Beiträge zur Lehre von den Temperaturempfindungen der menschlichen Haut. *Z. Biol.* **82**, 189–212.

REIN, F. H. (1925b). Über die Topographie der Warmempfindung. Beziehungen zwischen Innervation und receptorischen Endorganen. *Z. Biol.* **82**, 515–535.

REPIN, I. S. and KRATSKIN, I. L. (1967). Hypothalamic mechanisms of fever. *Neurosci. Transl.* **3**, 336–340, 1968 (translated from *Fiziol. Zh. SSSR I.M. Sechenova* **53**, 1206–1211).

RÉTHELYI, M., TREVINO, D. L. and PERL, E. R. (1979). Distribution of primary afferent fibers within the sacrococcygeal dorsal horn: an autoradiographic study. *J. comp. Neurol.* **185**, 603–622.

REVUSKY, S. H. (1966). Cold acclimatization in hairless mice measured by behavioural thermoregulation. *Psychonom. Sci.* **6**, 209–210.

REXED, B. (1952). The cytoarchitectonic organization of the spinal cord in the cat. *J. comp. Neurol.* **96**, 415–495.

RIEDEL, W. (1976). Warm receptors in the dorsal abdominal wall of the rabbit. *Pflügers Arch.* **361**, 205–206.

RIEDEL, W., SIAPLAURAS, G. and SIMON, E. (1973). Intra-abdominal thermosensitivity in the rabbit as compared with spinal thermosensitivity. *Pflügers Arch.* **340**, 59–70.

ROBERTS, M. F., WENGER, C. B., STOLWIJK, J. A. J. and NADEL, E. R. (1977). Skin blood flow and sweating changes following exercise training and heat acclimation. *J. appl. Physiol.* **43**, 133–137.

ROBERTSHAW, D. (1971). The evolution of thermoregulatory sweating in man and animals. *Int. J. Biometeorol.* **15**, 263–267.

ROBERTSHAW, D., TAYLOR, C. R. and MAZZIA, L. M. (1973). Sweating in primates: secretion by adrenal medulla during exercise. *Amer. J. Physiol.* **224**, 678–681.

ROBERTSON, R. T., LYNCH, G. S. and THOMPSON, R. F. (1973). Diencephalic distributions of ascending reticular systems. *Brain Res.* **55**, 309–322.

ROBINSON, S. (1949). Physiological adjustments to heat. *In* "Physiology of Heat Regulation and the Science of Clothing" (L. H. Newburgh, Ed.), pp. 193–231. Saunders, Philadelphia.

ROHLES, R. H. (1969). Preference for the thermal environment by the elderly. *Human Factors II*, 37–41.

ROIZEN, M. F. (1976). Biochemical mapping of noradrenergic projections of axons in the dorsal noradrenergic bundle. *Brain Res.* **104**, 384–389.

ROSENDORFF, C. and CRANSTON, W. I. (1968). Effects of salicylate on human temperature regulation. *Clin. Sci.* **35**, 81–100.

ROSENDORFF, C. and MOONEY, J. J. (1971). Central nervous system sites of action of a purified leucocyte pyrogen. *Amer. J. Physiol.* **220**, 597–603.

ROWELL, L. B., BRENGELMANN, G. L., DETRY, J.-M. R. and WYSS, C. (1971a). Venomotor responses to rapid changes in skin temperature in exercising man. *J. appl. Physiol.* **30**, 64–71.

ROWELL, L. B., BRENGELMANN, G. L., DETRY, J.-M. R. and WYSS, C. (1971b). Venomotor responses to local and remote thermal stimuli to skin in exercising man. *J. appl. Physiol.* **30**, 72–77.

RÓZSA, A. J. and KENSHALO, D. R. (1977). Bilateral spatial summation of cooling of symmetrical sites. *Percept. Psychophys.* **21**, 455–462.

RUDY, T. A. and WOLF, H. H. (1971). The effect of intrahypothalamically injected sympathomimetic amines on temperature regulation in the cat. *J. Pharmacol. Exp. Ther.* **179**, 218–235.

RUDY, T. A. and WOLF, H. H. (1972). Effect of intracerebral injections of carbamylcholine and acetylcholine on temperature regulation in the cat. *Brain Res.* **38**, 117–130.

RUWE, W. D. and MYERS, R. D. (1978). Dopamine in the hypothalamus of the cat: pharmacological characterization and push-pull perfusion analysis of sites mediating hypothermia. *Pharmacol. Biochem. Behav.* **9**, 65–80.

RYSER, G. and JÉQUIER, E. (1972). Study by direct calorimetry of thermal balance on the first day of life. *Europ. J. clin. Invest.* **2**, 176–187.

SALEH, M. A., HARO, P. J. and WINGET, C. M. (1977). Loss of circadian rhythmicity in body temperature and locomotor activity following suprachiasmatic lesions in the rat. *J. Interdiscipl. Cycle Res.* **8**, 341–346.

SALTIN, B. and GAGGE, A. P. (1971). Sweating and body temperatures during exercise. *Int. J. Biometeorol.* **15**, 189–194.

SALTIN, B., GAGGE, A. P. and STOLWIJK, J. A. J. (1968). Muscle temperature during submaximal exercise in man. *J. appl. Physiol.* **25**, 679–688.

SAMS, W. M. JR. and WINKELMANN, R. K. (1969). Temperature effects on isolated resistance vessels of skin and mesentery. *Amer. J. Physiol.* **216**, 112–116.

SAND, A. (1938). The function of the ampullae of Lorenzini, with some observations on the effect of temperature on sensory rhythms. *Proc. roy. Soc. Lond. B* **125**, 524–553.

SANS, K. (1949). Die Heißempfindung bei chemischer Reizung der äußeren Haut. *Inaug.-Diss.*, Heidelberg.

SATINOFF, E. (1964). Behavioral thermoregulation in response to local cooling of the rat brain. *Amer. J. Physiol.* **206**, 1389–1394.

SATINOFF, E. (1979). Drugs and thermoregulatory behavior. *In* "Body Temperature, Drug Effects and Therapeutic Implications" (P. Lomax and E. Schonbaum, Eds), pp. 151–181. Marcel Dekker Inc., New York.

SATINOFF, E. and HACKETT, E. R. (1977). Determination of set-point changes after intrahypothalamic injection of norepinephrine in rats. *In* "Drugs, Biogenic Amines and Body Temperature" (P. Lomax, E. Schönbaum and K. E. Cooper, Eds), pp. 87–95. Karger, Basel.

SATINOFF, E. and HENDERSEN, R. (1977). Thermoregulatory behavior. *In* "Handbook of Operant Behavior" (W. K. Honig and J. E. R. Staddon, Eds), pp. 153–173. Prentice-Hall, Englewood Cliffs, N.J.

SATINOFF, E. and RUTSTEIN, J. (1970). Behavioral thermoregulation in rats with anterior hypothalamic lesions. *J. comp. physiol. Psychol.* **71**, 77–82.

SATINOFF, E., McEWEN, G. N. JR and WILLIAMS, B. A. (1976a). Behavioral fever in newborn rabbits. *Science* **193**, 1139–1140.

SATINOFF, E., McEWEN, G. N. JR and WILLIAMS, B. A. (1976b). Behavioral and physiological responses to pyrogen in newborn rabbits. *Neurosci. Abstr.* **2**, 226.

SCHÄFER, K. and HENSEL, H. (1975). Peripheral components in short term habituation. *Pflügers Arch.* **359**, R98.

SCHÄFER, K., BRAUN, H. A. and HENSEL, H. (1978). Dependence of cold fibre response from blood calcium level in the cat. *Pflügers Arch.* **373**, R68.

SCHÄFER, K., BRAUN, H. A., BADE, H. and HENSEL, H. (1979). EDTA-induced burst discharge in cold fibres of the cat's nose. *Pflügers Arch.* **379**, R40.

SCHEVING, L. E., HALBERG, F. and PAULY, J. E. (1974). "Chronobiology". Thieme, Stuttgart.

SCHIFF, D., STERN, L. and LEDUC, J. (1966). Chemical thermogenesis in newborn infants: catecholamine excretion and the plasma non-esterified fatty acid response to cold exposure. *Pediatrics* **37**, 577–582.

SCHMIDT, K. L. (1975a). "Hyperthermie und Fieber. Wirkungen bei Mensch und Tier. Klinik, Pathologie, Immunologie. Wirkungen auf Arthritiden". Hippokrates, Stuttgart.

SCHMIDT, K. L. (1975b). Fieber, ein Abwehrmechanismus? *Dtsch. med. Wschr.* **100**, 1805–1808.

SCHMIDT, R. (1949). Die Empfindung der äußeren Haut bei Reizung durch Säuren, Laugen und Chloroform. Inaug.-Diss., Heidelberg.

SCHMIDT, W. (1976). Die Verarbeitung thermorezeptiver afferenter Signale im Trigeminuskern. Inaug.-Diss., Marburg.

SCHOENER, E. P. and WANG, S. C. (1975). Leukocytic pyrogen and sodium acetylsalicylate on hypothalamic neurons in the cat. Amer. J. Physiol. 229, 185–190.

SCHOLANDER, P. F., HAMMEL, H. T., LANGE ANDERSEN, K. and LØYNING, Y. (1958). Metabolic acclimation to cold in man. J. appl. Physiol. 12, 1–8.

SCHREINER, H. J. (1938). Das Wärmegefühl nach Calciuminjektionen. Inaug.-Diss., Düsseldorf.

SCHWENKENBECHER, A. (1908). Über Mentholvergiftung des Menschen. Münch. med. Wschr. 1495–1496.

SCHWENNICKE, H. P. and BRÜCK, K. (1976). Thermoregulatory modifications in the course of repeated moderate heat-exposures in man. Pflügers Arch. 365, R27.

SENAY, L. C. JR (1975). Plasma volumes and constituents of heat-exposed men before and after acclimatization. J. appl. Physiol. 38, 570–580.

SENAY, L. C. JR (1979). Temperature regulation and hypohydration: a singular view. J. appl. Physiol. 47, 1–7.

SENAY, L. C., MITCHELL, D. and WYNDHAM, C. H. (1976). Acclimatization in a hot, humid environment: body fluid adjustments. J. appl. Physiol. 40, 786–796.

SHAPIRO, C. M., MOORE, A. T., MITCHELL, D. and YODAIKEN, M. L. (1974). How well does man thermoregulate during sleep? Experientia (Basel) 30, 1279–1281.

SHEPHERD, J. T. and WEBB-PEPLOE, M. M. (1970). Cardiac output and blood flow distribution during work in heat. In "Physiological and Behavioral Temperature Regulation" (J. D. Hardy, A. P. Gagge and J. A. J. Stolwijk, Eds), pp. 237–253. Thomas, Springfield, Ill.

SIEGERT, R. (1977a). Der gemeinsame Mechanismus des Infektionsfiebers. Med. Klin. 72, 1787–1795.

SIEGERT, R. (1977b). Mechanismus des Virusfiebers. Dtsch. med. Wschr. 102, 204–208.

SIEGERT, R., PHILIPP-DORMSTROM, W. K., RADSAK, K. and MENZEL, H. (1976). Mechanism of fever induction in rabbits. Infect. Immun. 14, 1130–1137.

SILVERMAN, W. A., ZAMELIS, A., SINCLAIR, J. C. and AGATE, F. J. JR (1964). Warm nape of the newborn. Pediatrics 33, 984–986.

SIMON, E. (1971). Regional differentiation of vasomotor activity underlying thermoregulatory adjustments of blood flow. Int. J. Biometeorol. 15, 219–224.

SIMON, E. (1972). Temperature signals from skin and spinal cord converging on spinothalamic neurons. Pflügers Arch. 337, 323–332.

SIMON, E. (1974). Temperature regulation: The spinal cord as a site of extrahypothalamic thermoregulatory functions. *Rev. Physiol. Biochem. Pharmacol.* **71**, 1–76.

SIMON, E. and IRIKI, M. (1970). Ascending neurons of the spinal cord activated by cold. *Experientia (Basel)* **26**, 620–622.

SIMON, E. and IRIKI, M. (1971a). Ascending neurons highly sensitive to variations of spinal cord temperature. *J. Physiol. (Paris)* **63**, 415–417.

SIMON, E. and IRIKI, M. (1971b). Sensory transmission of spinal heat and cold sensitivity in ascending spinal neurons. *Pflügers Arch.* **328**, 103–120.

SIMON, E., RAUTENBERG, W., THAUER, R. and IRIKI, M. (1963). Auslösung thermoregulatorischer Reaktionen durch lokale Kühlung im Vertebralkanal. *Naturwissenschaften* **50**, 337.

SIMON, E., RAUTENBERG, W. and JESSEN, C. (1965). Initiation of shivering in unanaesthetized dogs by local cooling of the vertebral canal. *Experientia (Basel)* **21**, 476–477.

SIMPSON, C. W., RUWE, W. D. and MYERS, R. D. (1976). Characterization of prostaglandin sensitive sites in the monkey hypothalamus mediating hyperthermia. *In* "Drugs, Biogenic Amines and Body Temperature", pp. 142–144. Karger, Basel.

SKRAMLIK, E. VON (1937). Psychophysiologie der Tastsinne. *Arch. Psychol., Erg.* **4**, Parts 1 and 2.

SLEPCHUK, N. A. and IVANOV, K. P. (1972). On the thermo-sensitive interoceptors and their interaction with the hypothalamic thermo-sensitive structures. *Fiziol. Zh. SSSR* **58**, 1494–1498.

SMILES, K. A., ELIZONDO, R. S. and BARNEY, CH. C. (1976). Sweating responses during changes of hypothalamic temperature in the rhesus monkey. *J. appl. Physiol.* **40**, 653–657.

SMITH, J. B. and WILLIS, A. L. (1971). Aspirin selectively inhibits prostaglandin production in human platelets. *Nature (New Biol.)* **231**, 235.

SMITH, R. E. and HORWITZ, B. A. (1969). Brown fat and thermogenesis. *Physiol. Rev.* **49**, 330–425.

SMITH, R. I., PLATOU, E. S. and GOOD, R. A. (1956). Septicemia of the newborn. Current status of the problem. *Pediatrics* **17**, 549–575.

SNELLEN, J. W. (1972). Set point and exercise. *In* "Essays on Temperature Regulation" (J. Bligh and R. E. Moore, Eds), pp. 139–148. North-Holland Publ. Comp., Amsterdam and London.

SQUIRES, R. D. and JACOBSON, F. H. (1968). Chronic deficits of temperature regulation produced in cats by preoptic lesions. *Amer. J. Physiol.* **214**, 549–560.

STARY, Z. (1925). Über Erregung der Wärmenerven durch Pharmaka. *Naunyn-Schmiedeberg's Arch. exp. Path. Pharmak.* **105**, 76–87.

STEELE, R. E. and WEKSTEIN, D. R. (1972). Influence of thyroid hormone on homeothermic development of the rat. *Amer. J. Physiol.* **222**, 1528–1533.

STEELE, R. E. and WEKSTEIN, D. R. (1973). Effects of thyroxine on calorigenic response of the newborn rat to norepinephrine. *Amer. J. Physiol.* **224**, 979–984.

STEPHAN, F. K. and NUNEZ, A. A. (1977). Elimination of circadian rhythms in drinking, activity, sleep, and temperature by isolation of the suprachiasmatic nuclei. *Behav. Biol.* **20**, 1–16.

STERN, L., LEES, M. H. and LEDUC, J. (1965). Environmental temperature, oxygen consumption and catecholamine excretion in newborn infants. *Pediatrics* **36**, 367–373.

STEVENS, J. C. and STEVENS, S. S. (1960). Warmth and cold: dynamics of sensory intensity. *J. exp. Psychol.* **60**, 183–192.

STEVENS, J. C., ADAIR, E. R. and MARKS, L. E. (1970). Pain, discomfort, and warmth as functions of thermal intensity. *In* "Physiological and Behavioral Temperature Regulation" (J. D. Hardy, A. P. Gagge and J. A. J. Stolwijk, Eds), pp. 892–904. Thomas, Springfield, Ill.

STEVENS, J. C., MARKS, L. E. and SIMONSON, D. C. (1974). Regional sensitivity and spatial summation in the warmth sense. *Physiol. Behav.* **13**, 825–836.

STEVENS, S. S. (1958). Measurement and man. *Science* **127**, 383–389.

STEVENS, S. S. (1960). The psychophysics of sensory function. *Amer. Scientist* **48**, 226–253.

STEVENS, S. S. (1970). Neural events and the psychophysical law. *Science* **170**, 1043–1050.

STITT, J. T. (1973). Protaglandin E_1 fever induced in rabbits. *J. Physiol. (Lond.)* **232**, 163–179.

STITT, J. T. and HARDY, J. D. (1975). Microelectrophoresis of PGE_1 onto single units in the rabbit hypothalamus. *Amer. J. Physiol.* **229**, 240–245.

STITT, J. T., ADAIR, E. R., NADEL, E. R. and STOLWIJK, J. A. J. (1971). The relation between behavior and physiology in the thermoregulatory response of the squirrel monkey. *J. Physiol. (Paris)* **63**, 424–427.

STITT, J. T., HARDY, J. D. and STOLWIJK, J. A. J. (1974). PGE_1 fever: its effect on thermoregulation at different low ambient temperatures. *Amer. J. Physiol.* **227**, 622–629.

STOLWIJK, J. A. J. (1979). Physiological responses and thermal comfort in changing environmental temperature and humidity. *In* "Indoor Climate" (P. O. Fanger and O. Valbjørn, Eds), pp. 491–505. Danish Building Res. Inst., Copenhagen.

STOLWIJK, J. A. J. and HARDY, J. D. (1966). Temperature regulation in man. *Pflügers Arch. ges. Physiol.* **291**, 129–162.

STOLWIJK, J. A. J. and HARDY, J. D. (1977). Control of body temperature. *In* "Handbook of Physiology" (D. H. K. Lee, Ed.), Sect. 9, pp. 45–68. Amer. Physiol. Soc., Bethesda, Md.

STOLWIJK, J. A. J. and WEXLER, I. (1971). Peripheral nerve activity in response to heating the cat's skin. *J. Physiol. (Lond.)* **214**, 377–392.

STOLWIJK, J. A. J., SALTIN, B. and GAGGE, A. P. (1968). Physiological factors associated with sweating during exercise. *Aerospace Med.* **39**, 1101–1105.

STOLWIJK, J. A. J., NADEL, E. R., MITCHELL, J. W. and SALTIN, B. (1971). Modification of central sweating drive at the periphery. *Int. J. Biometeorol.* **15**, 268–272.

STREMPEL, H. (1978). Adaptive Modifikationen des Kälteschmerzes. II. Mitteilung: Langzeitversuche mit 24-stündigen Intervallen. *Europ. J. appl. Physiol.* **38**, 17–24.

STREMPEL, H. and TÄNDLER, P. (1976). Über die Bedeutung des Intervalles bei der Adaptation an serielle Kältereize. *Z. Phys. Med.* **6**, 16–17.

STRUGHOLD, H. and PORZ, R. (1931). Die Dichte der Kaltpunkte auf der Haut des menschlichen Körpers. *Z. Biol.* **91**, 563–571.

SULYOK, E., JÉQUIER, E. and PROD'HOM, L. S. (1973). Thermal balance of the newborn infant in a heat-gaining environment. *Pediatr. Res.* **7**, 888–900.

SUMINO, R., DUBNER, R. and STARKMAN, S. (1973). Responses of small myelinated 'warm' fibers to noxious heat applied to the monkey's face. *Brain Res.* **62**, 260–263.

SZELÉNYI, Z., ZEISBERGER, E. and BRÜCK, K. (1976). Effects of electrical stimulation in the lower brainstem on temperature regulation in the unanaesthetized guinea-pig. *Pflügers Arch.* **364**, 123–127.

SZELÉNYI, Z., ZEISBERGER, E. and BRÜCK, K. (1977). A hypothalamic alpha-adrenergic mechanism mediating the thermogenic response to electrical stimulation of the lower brainstem in the guinea pig. *Pflügers Arch.* **370**, 19–23.

TAM, H.-S., DARLING, R. C., CHEH, H.-Y. and DOWNEY, J. A. (1978). Sweating response: a means of evaluating the set-point theory during exercise. *J. appl. Physiol.* **45**, 451–458.

TAPPER, D. N., BROWN, P. B. and MORAFF, H. (1973). Functional organization of the cat's dorsal horn: connectivity of myelinated fiber systems of hairy skin. *J. Neurophysiol.* **36**, 817–826.

TEMPLETON, J. R. (1970). Reptiles. *In* "Comparative Physiology of Thermoregulation" (G. C. Whittow, Ed.), Vol. 1, pp. 167–221. Academic Press, New York and London.

THAUER, R. (1939). Der Mechanismus der Wärmeregulation. *Ergebn. Physiol.* **41**, 607–805.

THAUER, R. (1970). Thermosensitivity of the spinal cord. *In* "Physiological and Behavioral Temperature Regulation" (J. D. Hardy, A. P. Gagge and J. A. J. Stolwijk, Eds), pp. 472–492. Thomas, Springfield, Ill.

THOMPSON, F. J. and BARNES, C. D. (1969). Evidence for thermosensitive receptors in the femoral vein. *Fed. Proc.* **28**, 722.

THUNBERG, T. (1901). Untersuchungen über die relative Tiefenlage der kälte-, wärme- und schmerzperzipierenden Nervenenden in der Haut und über das Verhältnis der Kältenervenenden gegenüber Wärmereizen. *Skand. Arch. Physiol. (Berl. u. Lpz.)* **11**, 382–435.

TIMBAL, J., COLIN, J. and BOUTELIER, C. (1970). Étude de la phase transitoire de la sudation chez l'homme. *Pflügers Arch.* **318**, 305–314.

TIMBAL, J., COLIN, J. and BOUTELIER, C. (1971). Étude de la sudation de l'homme en régime transitoire. *J. Physiol. (Paris)* **63**, 442–445.

TOREBJÖRK, H. E. (1974). Afferent C units responding to mechanical, thermal and chemical stimuli in human non-glabrous skin. *Acta physiol. scand.* **92**, 374–390.

TOREBJÖRK, H. E. and HALLIN, R. G. (1972). Activity in C fibres correlated to perception in man. *In* "Cervical Pain" (C. Hirsch and Y. Zotterman, Eds), pp. 171–178. Pergamon Press, Oxford.

TOREBJÖRK, H. E. and HALLIN, R. G. (1974). Identification of afferent C units in intact human skin nerves. *Brain Res.* **67**, 387–403.

TOREBJÖRK, H. E. and HALLIN, R. G. (1976). Skin receptors supplied by unmyelinated (C) fibers in man. *In* "Sensory Functions of the Skin in Primates" (Y. Zotterman, Ed.), pp. 475–485. Pergamon Press, Oxford.

TOTEL, G. L., JOHNSON, R. E., FAY, F. A., GOLDSTEIN, J. A. and SCHICK, J. (1971). Experimental hyperthermia in traumatic quadriplegia. *Int. J. Biometeorol.* **15**, 346–355.

TRAUTMANN, A. and HILL, H. (1949). Temperaturmessungen im Pansen und Labmagen des Wiederkäuers (Ziege). *Pflügers Arch.* **252**, 30–39.

UNGERSTEDT, U. (1971). Stereotaxic mapping of the monoamine pathways in the rat brain. *Acta physiol. scand.* Suppl. 367.

VANE, J. R. (1971). Inhibition of prostaglandin synthesis as a mechanism of action for aspirin-like drugs. *Nature New Biol.* **231**, 232–235.

VANHOUTTE, P. and LEUSEN, I. (1969). The reactivity of isolated venous preparations to electrical stimulation. *Pflügers Arch.* **306**, 341–353.

VANHOUTTE, P. M. and SHEPHERD, J. T. (1969). Activity and thermosensitivity of canine cutaneous veins after inhibition of monoamine oxidase and catechol-o-methyl transferase. *Circulat. Res.* **25**, 607–616.

VANHOUTTE, P. M. and SHEPHERD, J. T. (1970). Effect of cooling on beta-receptor mechanisms in isolated cutaneous veins of the dog. *Microvascul. Res.* **2**, 454–461.

VANHOUTTE, P. M. and SHEPHERD, J. T. (1971). Thermosensitivity and veins. *J. Physiol. (Paris)* **63**, 449–451.

VEALE, W. L. and COOPER, K. E. (1974). Prostaglandin in cerebrospinal fluid following perfusion of hypothalamic tissue. *J. appl. Physiol.* **37**, 942–945.

VEALE, W. L. and COOPER, K. E. (1975). Comparison of sites of action of prostaglandin E and leucocyte pyrogen in brain. *In* "Temperature Regulation and Drug Action" (J. Lomax, E. Schönbaum and J. Jacob, Eds), pp. 218–226. Kargel, Basel.

VEALE, W. L. and WHISHAW, I. Q. (1976). Body temperature responses at different ambient temperatures following injections of prostaglandin E_1 and noradrenaline into the brain. *Pharmacol. Biochem. Behav.* **4**, 143–150.

VENDRIK, A. J. H. (1959). The regulation of body temperature in man. *Ned. T. Geneesk.* **103**, 240–244.

VENDRIK, A. J. H. and VOS, J. J. (1958). Comparison of the stimulation of the warmth sense organ by microwave and infrared. *J. appl. Physiol.* **13**, 435–444.

VILLABLANCA, J. and MYERS, R. D. (1965). Fever produced by microinjection of typhoid vaccine into hypothalamus of cats. *Amer. J. Physiol.* **208**, 703–707.

WAGNER, J. A., ROBINSON, S., TZANKOFF, S. P. and MARINO, R. P. (1972). Heat tolerance and acclimatization to work in the heat in relation to age. *J. appl. Physiol.* **33,** 616–622.

WAGNER, J. A., ROBINSON, S. and MARINO, R. P. (1974). Age and temperature regulation of humans in neutral and cold environments. *J. appl. Physiol.* **37,** 562–565.

WALLER, M. B., MYERS, R. D. and MARTIN, G. E. (1976). Thermoregulatory deficits in the monkey produced by 5,6-dihydroxytryptamine injected into the hypothalamus. *Neuropharmacology* **15,** 61–68.

WANG, L. C. H. and HUDSON, J. W. (Eds) (1978). "Strategies in Cold: Natural Torpidity and Thermogenesis". Academic Press, New York, San Francisco and London.

WATTS, A. J. (1972). Hypothermia in the aged: a study of the role of cold-sensitivity. *Environ. Res.* **5,** 119–126.

WEAVER, M. E. and INGRAM, D. L. (1969). Morphological changes in swine associated with environmental temperature. *Ecology* **50,** 710–713.

WEBB, P. (1970). Thermoregulation in actively cooled working men. *In* "Physiological and Behavioral Temperature Regulation" (J. D. Hardy, A. P. Gagge and J. A. J. Stolwijk, Eds), pp. 756–774. Thomas, Springfield, Ill.

WEBB, P., ANNIS, J. F. and TRAUTMAN, S. J. JR (1978). Heat flow regulation. *In* "New Trends in Thermal Physiology" (Y. Houdas and J. D. Guieu, Eds), pp. 29–32. Masson, Paris.

WEBB-PEPLOE, M. M. (1969a). Effect of changes in central body temperature on capacity elements of limb and spleen. *Amer. J. Physiol.* **216,** 643–646.

WEBB-PEPLOE, M. M. (1969b). Cutaneous venoconstrictor response to local cooling in the dog. Unexplained by inhibition of neuronal re-uptake of norepinephrine. *Circulat. Res.* **24,** 607–615.

WEBB-PEPLOE, M. M. and SHEPHERD, J. T. (1968a). Responses of the superficial limb veins of the dog to changes in temperature. *Circulat. Res.* **22,** 737–746.

WEBB-PEPLOE, M. M. and SHEPHERD, J. T. (1968b). Response of dogs' cutaneous veins to local and central temperature changes. *Circulat. Res.* **23,** 693–699.

WEBB-PEPLOE, M. M. and SHEPHERD, J. T. (1968c). Peripheral mechanism involved in response of dogs' cutaneous veins to local temperature change. *Circulat. Res.* **23,** 701–708.

WEBER, E. H. (1846). Der Tastsinn und das Gemeingefühl. *In* "Wagners Handwörterbuch der Physiologie", Vol. 3/2, pp. 481–588. Vieweg and Sohn, Braunschweig.

WEDDELL, G. and MILLER, S. (1962). Cutaneous sensibility. *Ann. Rev. Physiol.* **24,** 199–222.

WEDDELL, G., PALMER, E. and PALLIE, W. (1955). Nerve endings in mammalian skin. *Biol. Rev.* **30,** 159–195.

WEISS, B. (1957). Pantothenic acid deprivation and thermal behavior of the rat. *Amer. J. clin. Nutr.* **5,** 125–128.

WEISS, B. L. and AGHAJANIAN, G. K. (1971). Activation of brain serotonin metabolism by heat: role of midbrain raphe neurons. *Brain Res.* **26,** 37–48.

WEISS, B., LATIES, V. G. and WEISS, A. B. (1967). Behavioral thermoregulation by cats with pyrogen-induced fever. *Arch. int. Pharmacodyn. Ther.* **165,** 467–475.

WELLS, CH. L. and BUSKIRK, E. R. (1971). Limb sweating rates overlying active and nonactive muscle tissue. *J. appl. Physiol.* **31,** 858–868.

WENDLANDT, S. (1972). Some factors involved in the control of body temperature and the action of pyrogens and pharmacologically active substances in modifying temperature regulation. Ph.D. Thesis, University of London.

WENGER, C. B., ROBERTS, M. F., NADEL, E. R. and STOLWIJK, J. A. J. (1975a). Thermoregulatory control of finger blood flow. *J. appl. Physiol.* **38,** 1078–1082.

WENGER, C. B., ROBERTS, M. F., STOLWIJK, J. A. J. and NADEL, E. R. (1975b). Forearm blood flow during body temperature transients produced by leg exercise. *J. appl. Physiol.* **38,** 58–63.

WERNER, J. (1975). Zur Temperaturregelung des menschlichen Körpers. Ein mathematisches Modell mit verteilten Parametern und ortsabhängigen Variablen. *Biol. Cybernetics* **17,** 53–63.

WERNER, J. (1977a). Mathematical treatment of structure and function of the human thermoregulatory system. *Biol. Cybernetics* **25,** 93–101.

WERNER, J. (1977b). Influences of local and global temperature stimuli on the Lewis-reaction. *Pflügers Arch.* **367,** 291–294.

WERNER, J. (1978). A contribution to the problem of set-point of human temperature regulation. *In* "New Trends in Thermal Physiology" (Y. Houdas and J. D. Guieu, Eds), pp. 26–28. Masson, Paris.

WERNER, J., SCHINGNITZ, G. and HENSEL, H. (1980). Influence of cold adaptation on the activity of thermoreactive neurons in midbrain and thalamus of the rat. *Pflügers Arch.* **384,** R25.

WHITTOW, G. C. (Ed.) (1971). "Comparative Physiology of Thermoregulation", Vol. 2. Academic Press, New York and London.

WINSLOW, C. E. A. and HERRINGTON, L. P. (1949). "Temperature and Human Life". Princeton Univ. Press, Princeton.

WISSLER, E. H. (1966). A mathematical model of the human thermal system. *Chem. Engn. Med.* **62,** 66–78.

WIT, A. and WANG, S. C. (1968a). Temperature-sensitive neurons in preoptic/anterior hypothalamic region: effects of increasing ambient temperature. *Amer. J. Physiol.* **215,** 1151–1159.

WIT, A. and WANG, S. C. (1968b). Temperature-sensitive neurons in preoptic/anterior hypothalamic region: actions of pyrogens and acetylsalicylate. *Amer. J. Physiol.* **215,** 1160–1169.

WOOD, J. D. (1970). Electrical activity from single neurons in Auerbach's plexus. *Amer. J. Physiol.* **219,** 159–169.

WÜNNENBERG, W. and BRÜCK, K. (1968a). Zur Funktionsweise thermoreceptiver Strukturen im Cervicalmark des Meerschweinchens. *Pflügers Arch. ges. Physiol.* **299,** 1–10.

WÜNNENBERG, W. and BRÜCK, K. (1968b). Single unit activity evoked by thermal stimulation of the cervical spinal cord in the guinea pig. *Nature* (*Lond.*) **218**, 1268–1269.

WÜNNENBERG, W. and BRÜCK, K. (1970). Studies on the ascending pathways from the thermosensitive region of the spinal cord. *Pflügers Arch.* **321**, 233–241.

WÜNNENBERG, W. and HARDY, J. D. (1972). Response of single units of the posterior hypothalamus to thermal stimulation. *J. appl. Physiol.* **33**, 547–552.

WÜNNENBERG, W., MERKER, G. and SPEULDA, E. (1978). Thermosensitivity of preoptic neurons and hypothalamic integrative function in hibernators and nonhibernators. *In* "Strategies in Cold: Natural Torpidity and Thermogenesis" (L. Wang and J. W. Hudson, Eds), pp. 267–297. Academic Press, New York.

WURSTER, R. (1968). Beeinflussung der Trommelfell und Mundtemperatur durch Anderung der Kopfhauttemperatur. *Pflügers Arch. ges. Physiol.* **300**, R47.

WURSTER, R. D. and MCCOOK, R. D. (1969). Influence of rate of change in skin temperature on sweating. *J. appl. Physiol.* **27**, 237–240.

WURSTER, R. D., HASSLER, C. R., MCCOOK, R. D. and RANDALL, W. C. (1969). Reversal in patterns of sweat recruitment. *J. appl. Physiol.* **26**, 89–94.

WYNDHAM, C. H. (1966). Role of skin and core temperature in man's temperature regulation. *J. appl. Physiol.* **20**, 31–36.

WYNDHAM, C. H. (1973). The physiology of exercise under heat stress. *Ann. Rev. Physiol.* **35**, 193–220.

WYNDHAM, C. H. and ATKINS, A. R. (1968). A physiological scheme and mathematical model of temperature regulation in man. *Pflügers Arch.* **303**, 14–30.

WYNDHAM, C. H., MORRISON, J. F., WILLIAMS, C. G., BREDELL, G. A. G., PETER, J., RAHDEN, M. J. E., VON, HOLDSWORTH, L. D., GRAAN, C. H. VAN, RENSBURG, A. J. VAN and MUNRO, A. (1964). Physiological reactions to cold of Caucasian females. *J. appl. Physiol.* **19**, 877–880.

WYNDHAM, C. H., ROGERS, G. G., SENAY, L. C. and MITCHELL, D. (1976). Acclimatization in a hot, humid environment: cardiavascular adjustments. *J. appl. Physiol.* **40**, 779–785.

WYSS, C. R., BRENGELMANN, G. L., JOHNSON, J. M., ROWELL, L. B. and NIEDERBERGER, M. (1974). Control of skin blood flow, sweating and heart-rate: role of skin vs. core temperature. *J. appl. Physiol.* **36**, 726–733.

WYSS, C. R., BRENGELMANN, G. L., JOHNSON, J. M., ROWELL, L. B. and SILVERSTEIN, D. (1975). Altered control of skin blood flow at high skin and core temperatures. *J. appl. Physiol.* **38**, 839–845.

YAMADA, M., OKADA, G., OTANI, T. and FUKUSHIMA, Y. (1976). Thermal sensibility of the receptor in the knee joint. *Yonago Acta med.* **20**, 66–73.

YAMASHITA, H. (1977). Effect of baro- and chemoreceptor activation on supraoptic nuclei neurons in the hypothalamus. *Brain Res.* **126**, 551–556.

ZEISBERGER, E. (1978). Role of a central catecholaminergic system in the thermoregulatory threshold deviation. *In* "New Trends in Thermal Physiology" (Y. Houdas and J. D. Guieu, Eds), pp. 22–25. Masson, Paris, New York, Barcelona and Milan.

ZEISBERGER, E., BERGER-GAUDE, E. and BRÜCK, K. (1977). Drug effects on hypothalamic structures participating in the control of thermogenesis. *In* "Depressed Metabolism and Cold Thermogenesis" (L. Janský, Ed.), pp. 182–187. Charles University, Prague.

ZEISBERGER, E. and BRÜCK, K. (1971a). Central effects of noradrenaline on the control of body temperature in the guinea pig. *Pflügers Arch.* **322**, 152–166.

ZEISBERGER, E. and BRÜCK, K. (1971b). Effect of intrahypothalamic noradrenaline-injection on the threshold temperatures for shivering and non-shivering thermogenesis. *J. Physiol. (Paris)* **63**, 464–467.

ZEISBERGER, E. and BRÜCK, K. (1973). Effects of intrahypothalamically injected noradrenergic and cholinergic agents on thermoregulatory responses. *In* "The Pharmacology of Thermoregulation" (E. Schönbaum and P. Lomax, Eds), pp. 232–243. Karger, Basel.

ZEISBERGER, E. and BRÜCK, K. (1976). Alteration of shivering threshold in cold- and warm-adapted guinea pigs following intrahypothalamic injection of noradrenaline and of an adrenergic alpha-blocking agent. *Pflügers Arch.* **362**, 113–119.

ZEISBERGER, E., SZELÉNYI, Z. and BRÜCK, K. (1977). Role of catecholaminergic brain stem pathways in thermoadaptive alteration of shivering threshold. *In* "Drugs, Biogenic Amines and Body Temperature", pp. 5–12. Karger, Basel.

ZELIS, R. and MASON, D. T. (1969). Comparison of the reflex reactivity of skin and muscle veins in the human forearm. *J. clin. Invest.* **48**, 1870–1877.

ZIMMERMANN, M. (1977). Encoding in dorsal horn interneurons receiving noxious and non-noxious afferents. *J. Physiol. (Paris)* **73**, 221–232.

ZITNIK, R. S., AMBROSIONI, E. and SHEPHERD, J. T. (1971). Effect of temperature on cutaneous venomotor reflexes in man. *J. appl. Physiol.* **31**, 507–512.

ZOTTERMAN, Y. (1935). Action potentials in the glossopharyngeal nerve and in the chorda tympani. *Skand. Arch. Physiol.* **72**, 73–77.

ZOTTERMAN, Y. (1936). Specific action potentials in the lingual nerve of cat. *Skand. Arch. Physiol.* **75**, 105–119.

ZOTTERMAN, Y. (1953). Special senses: thermal receptors. *Ann. Rev. Physiol.* **15**, 357–372.

Subject Index

MONOGRAPHS OF THE PHYSIOLOGICAL SOCIETY

Published by EDWARD ARNOLD

Published by CAMBRIDGE UNIVERSITY PRESS

Published by ACADEMIC PRESS

Volumes marked * are now out of print